Gardens and
Gardening in the Chesapeake

▲

Gardens
and Gardening
in the Chesapeake

1700–1805

BARBARA WELLS SARUDY

Nina so excited to begin to meet you as you begin to bring orchards back to Baltimore & beyond

Barbara Wells Sarudy

The Johns Hopkins University Press
Baltimore & London

This book has been brought to publication with the generous assistance of the Laurence Hall Fowler Fund.

The Johns Hopkins University Press
2715 North Charles Street
Baltimore, Maryland 21218-4363
The Johns Hopkins Press Ltd., London

Library of Congress Cataloging-in-Publication Data will be found at the end of this book.
A catalog record for this book is available from the British Library.

ISBN 0-8018-5823-2

Contents

Color plates follow page 114

Preface

Every now and then, one of my historian friends will lose control and blurt out, "What in the world are you doing studying gardens?" This work is my answer. In 1981, I began to do volunteer research in the colonial manuscripts at the Maryland State Archives. My task was to plow through hundreds of inventories, each one listing a person's belongings at death. The historians I was assisting were using these court records to examine the lives of early Marylanders. Often, colonial inventory clerks catalogued furnishings room by room, and as a result, scholars have gotten a good idea of how people lived inside their homes. Court and church records have helped us reconstruct how long and well these colonials lived and how their lives changed over time.

Soon it struck me how little appeared in these documents about the land immediately surrounding colonists' houses—the gardens and grounds. While houses protected the furnishings, so that some pieces exist even today, nature quickly reclaimed the gardens. But why should we worry about lost gardens at all?

The eighteenth century was the culmination of thousands of years of agrarian society. The nineteenth century would bring the Industrial Revolution to America. Until then, most societies based their economies on the raising and trading of crops, so nature was always in control. People measured the work day by the rising and setting of the sun, and one hailstorm or flood could ruin a year's work. Everyday life was an ongoing struggle against nature.

Historically when people have been able to raise enough crops and food to sustain a comfortable life, they have challenged nature even further by turning their outdoor environment into a living art form, a pleasure garden. Most societies have even given the garden religious significance.

A garden is a balance between measured, human control on one hand and wild, mystical nature on the other. It is the place where humans attempt to create their particular vision of an idealized order of nature and culture. A garden is not just the opposition of unpredictable nature and organized society; it is the mediating space between them. Human intellect, intuition, nurture, and spirit meld together in a garden. Since culture shapes both the form and the meaning of a

garden at a particular place and time, it seemed to me that if historians believed it significant to learn how agrarian colonials organized life inside their houses, it was at least as important to study the grounds they designed immediately around their homes. What did early American gardens look like? What social interactions took place in and about gardens? What did these gardens symbolize to the people who planned and maintained them?

Finding answers to these questions posed a challenge. Inventories and court records, so rich in information about the inside of people's homes, rarely gave any clues about their outdoor environments. Historians do not dig in the soil; they dig in records. No wonder they tend to leave garden history to archaeologists, landscape architects, and art historians.

The gardens of a few colonial gentry, such as George Washington and Thomas Jefferson, have been easier for historians to deal with because of these gardeners' careful record keeping, but we had known little about smaller gardens, like those of shopkeepers and craftspeople, whose numbers grew in colonial towns during the second half of the eighteenth century. In fact, the small and tidy geometric gardens surrounding the re-created craftsmen's homes at Colonial Williamsburg are often criticized by historians who doubt that busy, semieducated artisans could have created intricate, time-consuming gardens. The well-documented gardens of Washington and Jefferson were not typical Chesapeake pleasure gardens, even of the gentry.

As I dug, little by little, leads began to appear. Inventory clerks had recorded a few gardening books and garden ornaments. From time to time booksellers and libraries had listed garden books in their newspaper advertisements, and house-for-sale notices sometimes described the surrounding grounds. Contemporary paintings, maps, and drawings offered insights as well. Colonial letters, journals, and catalogues occasionally provided some garden clues. Soon, the bits and pieces of information about Chesapeake gardeners of all sorts began to fall into patterns.

Almost immediately it became apparent that early American garden history did not parallel British garden history of the period. Colonial Americans did not rush to adopt the "natural" pleasure grounds craze that was changing the British countryside at the time.[1] Why? By the eighteenth century in Britain, aristocratic families had farmed grand estates of a thousand acres or more for centuries. The mid-1700s found famous landscape designers traveling from one great estate to the next redesigning the face of a land already tightly sectioned off into hedged parcels. English gentry were escaping to their country estates and to smaller second homes and gardens on the Thames near Richmond, to avoid the corrupting forces of court politics and urban commerce.

In early America we find a wholly different situation. While the names Chesapeake gentry chose for their estates may have reflected the ideal of innocent rural retirement—Solitude, the Retreat, the Hermitage—they knew these plantations were serious business. These colonials were not just tending their pleasure gardens, they were actively managing the day-to-day struggles of carving out a comfortable life from the still untamed American countryside.[2] Visitors were well aware of the differences. "In America," a French traveler noted in the 1790s, "a very pretty country house corresponds only to a place moderately kept up on the outskirts of a large French city, and even then one will find in [America] neither the good taste . . . nor the comforts which make living in it a pleasure."[3]

Among early Americans, there was no great social gulf between the landed aristocrats and the mass of local gentry and shopkeepers. American gardens did not symbolize political disagreements between Whigs and Tories. British Americans generally shared a conservatism that undergirded the emerging new representative democracy. The new American government took inspiration from the Roman republic and drew much of its symbolism from classical sources.

Unlike Britain, America had an abundance of natural resources. British pleasure parks served as timber nurseries for the gentry when trees became scarce in the countryside. The colonial landscape still offered lush virgin woods. Early American landowners did not need forest keepers, they needed forest clearers.

The English gentry also designed and stocked their pleasure parks as protected private nesting retreats for wild fowl and game animals. Foresters patroled the "natural" gardens of eighteenth-century England. In the American countryside, deer, small mammals, and wild birds were plentiful, and people were free to hunt for game unencumbered by the English laws that allowed only the privileged to hunt and which punished poachers, sometimes with hanging.[4] English landowners were deadly serious about protecting game in their pleasure parks from not

only human but animal predators: English fox hunting was more than sport; it was war against an impudent animal that dared to feast on the gentry's game.

In eighteenth-century America wild game was abundant.[5] Having enough to eat, however, was not guaranteed. Most colonial families struggled to raise enough off the land to feed themselves and their workers and to have some extra to sell for some cash. While laboring simply to survive, some of these British Americans made earnest attempts at pleasure gardening. Their triumphs are revealing exactly because of these cultural and economic pressures and priorities. We can trace the

A Forrester

historical precedents for many of
their garden designs, but the motives behind American gardens of
this period often reflected a new
idealism, shared equality, and a
spirit of rebellion not common in
mother England.

.1. Forrest Keeper

I use the term *Chesapeake* loosely
here, referring to an area extending from Pennsylvania to Virginia.
Country landowners within this
region often built their houses on
a rise of ground, preferably on the
bank of a river or a bay, and then
carved the hills where their houses
sat into level garden areas con-
nected by sometimes steep turf ramps. Though it is true that gardeners up and
down the Atlantic included terraces in their garden designs, this penchant for terraces, slopes, and falls truly characterized mid-Atlantic taste. Those fortunate
enough to build on grounds falling toward the great Chesapeake Bay seemed particularly enamored with this style, which allowed the owner to manipulate both
access to and views from his house. The preference owed as much to control as to
art.

Many garden historians scour English garden history for models of eighteenth-
century colonial gardens. They pore over garden treatises and geometry texts hoping to stumble on some magical mathematical formula that American garden
builders followed. They rehash the easily accessible, well-documented but unique
garden efforts of Washington and Jefferson in an attempt to explain colonial gardening in general. And they read descriptions of contemporary American gardens by British travelers, who borrow terminology from grand English villas and
grounds to describe what they saw. Art historians warn them to discount pictorial
depictions of early American gardens and landscapes as copies from European
prints, and there is a little truth in all of it.

Thomas Jefferson, who had toured gardens in Europe and Britain, wrote to his
friend William Hamilton in 1806 that Hamilton's garden at the Woodlands, near
Philadelphia, was "the only rival which I have known in America to what may be
seen in England."[6] The difference was in the details, of course, just as one Englishman said when he compared Britain's Richmond on the Thames to America's
Richmond on the James in 1796: "The general landscapes from the two Rich-
mond-hills are so similar in their great features, that at first sight the likeness is

most striking. The detail of course must be extremely different. . . . The want of finish and neatness in the American landscape would first strike his eye."[7]

This study attempts to look at early American gardens planned by a broad range of people. It examines what these gardens looked like; who planted, supplied, maintained, and enjoyed them; and what these gardens meant to those people. This work searches for the reasons behind the differences in the details. In general, of course, the story of any garden at any time is a tale of the human attempt to deal with the awesome power of nature. And in the end, the mortal always loses, but the attempt is the story. Why early Americans struggled with this frustration is as compelling as the order they created out of the great American wilderness.

Thanks to two popular source books, *Old English Cuts and Illustrations* (New York: Dover Publications, 1970) and *1800 Woodcuts by Thomas Bewick and His School* (Dover, 1962 and 1990), this text is inhabited by people. Most images of Baltimore City houses and gardens are taken from the familiar Warner and Hanna "Plan of the City and Environs of Baltimore," drawn by Charles Varle in 1797, engraved by Francis Shallus, and published first in 1799 and the second edition in 1801 (available at the Maryland Historical Society).

Pleasure gardens in early America grew from the interactions and efforts of a variety of people at many levels of society, and I have tried to write this account of them for a diversity of readers.

Acknowledgments

J ohn Dixon Hunt, editor of the British *Journal of Garden History,* first published the seeds of this work as the July–September number of 1989, and I acknowledge my debt to him. Decorative arts historian Gregory R. Weidman first introduced me to the Maryland State Archives in 1981, and I am grateful to her for that. Historian and friend Ed Papenfuse has directed the work of the archives for years and has made it one of the best such repositories in the nation. It was in the archives that I met my idol, historian Lois Greene Carr, who knows practically everything about the colonial Chesapeake. Also entrenched in state government is the head of the Division of Historical and Cultural Programs, J. Rodney Little, an irreplaceable friend and champion. And filling that role closer to home is Judy Dobbs, Deputy Director of the Maryland Humanities Council.

I have learned more about eighteenth-century urban gardens from my hours of talking with Williamsburg gardener Terry Yemm than I have from anyone else. His practical perspective adds such depth to the abstract nature of studying documents. Over the years the staff of the Colonial Williamsburg Foundation have been particularly helpful to me. Historian Pat Gibbs has not only helped with research on women and gardens, she has housed me as well. Architectural historian Carl Lounsbury has spent hours assisting me with primary sources that refer to gardens and outbuildings. Historians Lorena Walsh and Linda Rowe are eager to share their knowledge of the land and its people as is landscape architect M. Kent Brinkley. Indispensable in my research has been my sponsor and friend Ron Hoffman, director of the Ohmohundro Institute for Early American History and Culture, and his assistant, and my friend, Sally Mason. Also willing to help whenever called on has been the past director of the institute, Thad Tate. I am grateful to them all.

Elsewhere in Virginia, Jefferson's vision flourishes at Monticello because of Peter Hatch, who directs the gardens and grounds. Each time I talk with him, I see the early American garden from a new and exciting perspective. Peter is a gardener with the soul of a poet. Thomas Jefferson would have appreciated the art of Peter Hatch. Peggy C. Newcomb is the force that runs the Center for Historic

Plants. She is a working gardener and a historian, a combination to envy. Years ago Mount Vernon's horticulturalist, Dean Norton, toured some of Maryland's terraced gardens with me. His insights were invaluable, and his knowledge of boxwoods is simply amazing.

The horticulturalist for Historic Annapolis is the elegant Lucy Dos Passos Coggins, who knows more about eighteenth-century Chesapeake Bay plants than anyone I know. She tirelessly hunts them, grows them, photographs them, and records references to them. And she is always willing to share her findings. Historic Annapolis, Incorporated, is also the home base of colonial historian and friend Jean Russo, whose knowledge and good humor are boundless. My favorite colleagues, Bea and Steve Hardy donate garden sources whenever they find them and are always just a phone call or e-mail away when I have a question.

Finally, I thank those I love whose support has been unfaltering—my parents, Helen and Wayne Wells; my husband, Richard; my children, Richard, Catherine, Christine, and Caroline; my physician Peter A. Holt; and my friend Peter J. Kim.

I
PLACES

A Craftsman's Garden

William Faris, the son of a London clockmaker, was brought to the colonies in 1728 at the age of six months by his recently widowed mother, and he scrambled all his life to make a respectable living. In Annapolis, he designed silver teapots and spoons, struggled to build a pianoforte, assembled tiny watches and towering tall clocks, kept an inn, pulled neighbors' teeth (and hung them on a string by his workbench), and annually contracted to wind the clocks at the state capitol and in the homes of the gentry.

In whatever spare time he could find, he gardened. Of course, he grew food for his wife, six children, and inn patrons; but surprisingly he also designed intricate flower beds, near the front of his lot, where his neighbors could admire them. Even though he seldom spelled a word the same way twice, he kept a diary, filled with his gardening triumphs and failures, for the last twelve years of his life—704 pages between 1792 and 1804. The terraced gardens of the Chesapeake gentry are easier to learn about than the smaller town gardens of the tradespeople, whose numbers were growing during the latter half of the eighteenth century, but Faris's invaluable journal offers a rare opportunity to reconstruct the town garden of an early American artisan.[1]

When William Faris settled in Annapolis at mid-century, the capital was at the height of its political, economic, and social dominance in Maryland. In 1696, when Royal Governor Francis Nicholson designed the port city of Annapolis, he plotted the new capital to sit high up, defensively overlooking the surrounding waters. Two circles, one containing the statehouse and the other the Anglican church, dominated the town's baroque design. Streets radiated from both circles. Visitors declared the town's prospects to be "bold and high," the roads leading out "like rays from a centre."[2]

Annapolis was designed as a stage for the social and political affairs of the province of Maryland. During the second half of the century, Chesapeake gardeners, gentry and artisans alike, designed the grounds surrounding their homes as their personal stages, on which they presented themselves to those passing by.[3]

William Faris's house sat on one of the streets radiating out of Church Circle,

only a few hundred feet from the church. In the spring of 1804, Faris's private Eden sat behind a bright-red wooden gate at the front entrance to his grounds. Eighteenth-century Maryland gateways, smaller and simpler than their European precedents, were still intended to limit access to their owner's property. They also marked changes in personal roles as people crossed from one side to the other. Outside that gate, Faris was a tired 75-year-old clockmaker, with thinning hair pulled back into a queue and covered with a familiar frayed hat, who gossiped too much and drank gin too freely. But on the other side of the bright gate, the old man blossomed. Here was the world he had mastered for over forty years. The red gate opened in a recently built stone wall that stretched 75 feet from the side of Faris's house to his neighbor's property line and ran along the edge of the town's busiest trade street. The craftsman's 36-foot-wide combination home, inn, and shop, "At the Sign of the Crown and Dial," sat directly on West Street. Like many other Chesapeake town gardens, Faris's began in a side lot and stretched to the rear. The adjoining new stonework replaced an old wooden picket fence.[4]

Behind the wall, the clockmaker's grounds were enclosed by picket fences and ran back 366 feet to a sleepy rear street, where the lot widened to 200 feet. Wooden fences surrounded most eighteenth-century Maryland gardens, which were usually described in local newspaper property-for-sale ads as "well paled in." Chesapeake picket fences were almost invariably painted white but were of differing designs.

Faris and his neighbors felt that fences of one sort or another were an absolute necessity, to discourage uninvited human and animal visitors as well as to demarcate their property boundaries. Chesapeake gardeners could either buy their fence posts from local suppliers or employ "a couple of stout hands in mauling fence logs." Faris's neighbors Charles Carroll of Annapolis (1702–83) and his son Charles Carroll of Carrollton (1737–1832) used their slaves to produce garden pales. Fancy wooden paling constructed to emulate Chinese designs was being advertised for sale in the Chesapeake region by the late 1760s. Variety of design became important as many town governments demanded that every homeowner enclose his land.[5]

In the colonies, garden interlopers were not searching for game or timber, as in Britain; they were looking for the fruits of the gardener's labor or were simply accidental tourists. Livestock occasionally roamed the streets in early American towns, and tender garden plants did not stand a chance under their feet. Human garden intrusion was more focused. One night in 1792, Faris startled a thief in his garden and recorded that his subsequent flight "broke off the top of one of the pales." But the incident that really angered him was when a thief stole into his garden one dark night in 1803 to steal a dozen of his most prized possessions—his tulips.[6]

Tulips were the old man's obsession. At the height of their blooming, Faris would find himself engulfed in a flood of color. This artisan and innkeeper grew thousands of tulips each year; he counted 2339 in the spring of 1804.[7] Tulips were not the only bulb flowers that caught his fancy: in 1798, he planted 4000 narcis-

Dutch Tulips

sus bulbs, bought from a neighbor.[8] This tireless gardener's greatest pleasure was creating new varieties of tulips in nursery beds at the back of his property.[9]

Faris saw his tulips as symbols of the new nation as well as reflections of classical republican ideals. On the eve of July 4, 1801, exactly twenty-five years after the signing of the Declaration of Independence, Faris listed in his journal his tulip varieties by name. Namesakes included Presidents Washington and Madison and classical heros such as Cincinnatus.

Faris gardened for money as well as love. Each spring he invited his neighbors to view his tulips at the height of their glory. Admiring visitors would mark varieties that caught their eye with sticks notched with a personal code. When the tulips died back in June, Faris would dig up the bulbs near the notched sticks, and the admirers would return to buy them. The craftsman had plenty to spare.[10]

The ornamental garden beds the craftsman designed in the 1760s were akin in design, if not grandeur, to the more elegant geometric gardens that Chesapeake gentry were busy building about the same time. After he bought and enlarged his combination house and business, Faris hired an English indentured servant gardener, in 1765, to help him install the basic design of his gardens.[11] Just as in the gardens of most Chesapeake gentry, straight paths and walkways formed the skeleton of his garden. Faris's grounds were divided by both grass and composition walks separating boxwood-lined beds; such paths were essential for walking, cultivation, and defining the garden beds.

Designs for most Chesapeake gardens appeared to strive for uniformity in every part; exact levels, straight lines, parallels, squares, circles, and other geometrical figures were used to effect symmetry and proportion. Straight walks were everywhere, arranged parallel and crossing one another at regular intersections, as they connected spaces and led from scene to scene.[12]

Faris planned small geometric beds on his compact town property, where economy of scale was essential. These beds were planted with low-growing vegetables and brightly flowering plants within the boxwood borders that outlined and decorated the space even after the flower season was past.[13]

Faris kept the walkways that divided his garden beds in immaculate condition. This required constant maintenance. Faris's female slave, who was his regular gardening companion, was busy each spring and fall sweeping and raking the composition garden walks, which were some combination of gravel, crushed oyster shells, sand, and pulverized brick. Even old Faris himself, who often experienced crippling pain in his hips, spent days bending down to clean his gardens and walkways of stones, extraneous shells, weeds, and falling petals.[14]

Faris also criss-crossed his grounds with grass paths, lined with boxwood, that would be pleasant and cool to the feet; but the hard, slightly convex composition walkways allowed for quick water drainage and drier walking in wet weather. He

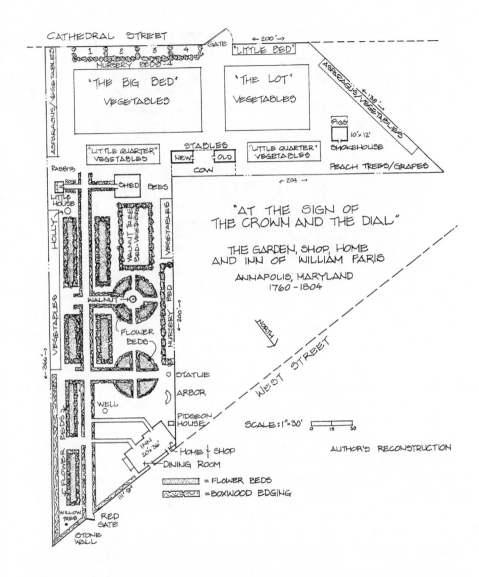

Map of William Faris's garden, Annapolis. William Faris used straight paths and walkways to form the skeleton of his gardens and grounds, which nearly filled a triangle bordered by Cathedral and West streets. Drawing by Susan Wirth.

paved the walk to the privy, which had to be used regardless of weather, with stones and crushed shells. Within the boxwood-lined spaces formed by the intersecting composition and grass walks, Faris planted all of his flower beds and some of his vegetable patches year after year.

Faris used boxwood for more than year-round definition of garden beds; box edging gave protection to seedlings and to newly picked-from greens expected to

produce more leaves for later rounds of harvesting. Faris and his garden helpers devoted days each spring to cutting the boxwood on his grounds. They were not just trimming the new green growth; they were also cutting the roots of each boxwood on the garden side as close to the plant as possible, so that the shallow roots would not rob the soil of the nutrients and moisture necessary for the other plants in the garden.[15]

Faris planned his grounds so that from the front of his property observers would see only the pleasure garden areas of his grounds, a host of geometric beds annually planted with flowers in the Dutch tradition. The beds were all bordered by grass walkways, adjoining property lines, or by one of the rectangular out-buildings on the property. He filled the rectangular beds on each side of the main grass walk with tuberoses, tulips, anemones, Chinese asters, crown imperials, globe amaranthus, and larkspur. The long composition walkway leading to the "neces-sary" was flanked by box-lined rectangular beds starring carefully trimmed holly trees surrounded by a supporting cast of tuberoses, white roses, India pinks, Chi-nese asters, tulips, hyacinths, and jonquils. Faris collected his holly trees from nearby woods and kept them trimmed in the shape of sugar cones.[16]

A narrow rectangular flower border next to the picket fence along an adjacent lot featured Job's tears, satin flowers, India pinks, snapdragons, tulips, and flow-ering beans that climbed the fence posts and trailed along the wooden rails. Faris planted one of his several nursery beds in the half of the garden nearer the house. There he grew the flowers to supply his various pleasure beds, propagated vast varieties of tulips and perennials, and heeled-in the boxwood cuttings he used to outline his garden beds. Not all of the craftsmen's flower beds were rectangular in shape. The area behind the house was dominated by a walnut tree. Nestled around its base was a circular bed divided into boxwood-lined quarters filled with tulips and bleeding hearts in May, followed by a succession of bright perennials throughout the summer months. Not far from the walnut tree, Faris planted a corresponding quartered circular bed also outlined with boxwood. The colorful circle overflowed with a profusion of polyanthus, tuberoses, wall flowers, India pinks, Chinese asters, hyacinths, jonquils, and tulips. In fact, wherever Faris planted flower beds, he included tulips. Sometimes, he even squeezed an errant tulip or two into his vegetable beds.

Usually, though, he separated his utility gardens from his ornamental areas, subscribing to the advice that the English garden writer William Lawson offered in his *New Orchard and Garden* in 1618: "Garden flowers shall suffer some disgrace, if among them you intermingle Onions, Parsnips, &c." Faris devoted the majority of his land to growing vegetables and fruits.

Sugar Loaves

Faris's occasional Annapolis neighbor, John Beale Bordley, gave growers advice on the kitchen garden, which he said should be an acre and a half for a small family like Faris's and four to five acres for a large one. Bordley said the kitchen garden should be exposed to the sun, not overshadowed with trees or buildings. He explained that the "soil should be of a pliable nature and easy to work; but by no means wet; and two feet, at least, deep." Bordley also advised that the kitchen garden should sit as "near the stables as possible, for the convenience of carrying dung.[17]

Walking toward the rear of Faris's property, the first utilitarian area one would be aware of was a vegetable bed along the left boundary. Then one would encounter a small rectangular plot Faris planted with vegetables every year, one of two called "little quarter," flanking the stables. There Faris grew unobtrusive vegetables and herbs that did not need much room to grow, including cabbages, carrots, peas, onions, thyme, spinach, curled savory, and several varieties of beans.

Although Faris almost always segregated his flower beds from his vegetable plots, he did not separate herbs from vegetables. In one of the "little quarters," Faris planted cabbages, asparagus, parsley, and Job's tears. After he built his new stable, he added an additional narrow rectangular bed, where he grew radishes, lettuce, nutmeg, and cherry peppers. On a border at the end of his new stable, which was visible from the main walkway leading to the rear of the lot, Faris occasionally grew a combination of flowers and vegetables: marigolds, lily of the valley, asters, balsam, anemones, and globe amaranthus nestled among bunch beans, spinach, radishes, and cherry tree seeds.

Not far from the new stable, the innkeeper maintained another rectangular vegetable patch dubbed "the walnut tree bed," where he grew beans, brussels sprouts, lettuce, kale, corn, and radishes. Faris diligently tended two separate asparagus plots near the back street, where he nudged a few more lettuce, cabbage, and spinach plants in between the tender green springtime shoots. A great portion of the vegetables Faris fed his family and guests came from a larger vegetable plot which he called simply "the garden" or the "big bed." Faris outlined even this large rectangular vegetable garden with exact rows of sage and rosemary, which he kept trimmed and orderly. The "big bed" lay close to the stables and the smokehouse at the rear of the property. There Faris planted peas, parsnips, corn, cabbage, cauliflower, radishes, beans, cucumbers, squash, cantaloupes, and watermelons.

The craftsman devoted the largest space at the rear of his grounds solely to kitchen gardening. He referred to this particular area as "the outer lot" or "the lot." Spreading plants like squash, musk melons, cucumbers, watermelons, and cantaloupes grew there. Faris occasionally scattered early crops of cabbages, carrots, greens, parsnips, radishes, brussels sprouts, and kale among the maturing vines; but he usually grew his compact vegetables in smaller patches, such as the narrow bed that ran down one side of his fenced property line, where he planted slender rows of small vegetables, including cabbages, lettuce, onions, brussels sprouts, spinach, and peas.

Faris used the picket fence along the back of his grounds to help define a set of four rectangular nursery beds for rearing fledgling tulips and boxwood cuttings. Even these nursery compartments he outlined with neatly trimmed ivy borders and boxwood. Near his bee house, Faris planted additional rows of peas, beans,

Inn-keeper

cabbage, kale, parsley, and cherry peppers. Not one to let any space go to waste, he squeezed a few more radishes, lettuce, cabbages, and parsnips into a narrow rectangular space under the window of his public dining room.

Atune to the times, in 1799 Faris added a wooden porch and steps to the back of his house, overlooking the garden area.[18] The Carrolls had added an elegant porch with stone columns to their Annapolis home when they remodeled their gardens in the 1770s, and many Baltimoreans and Philadelphians were also building porches or piazzas onto their homes during this period.[19] The addition of piazzas to Chesapeake homes in the last quarter of the eighteenth century coincided with the expansion of leisure time and the development of ornamental gardens. The simple geometric garden designs used in gardens of the period were seen to best advantage from a higher level, such as an upper terrace, second-story windows, or a porch. These prospects also allowed the homeowner and his guests a better vantage point from which to survey both the gardener's efforts at ordering nature around him and the surrounding countryside beyond.

Near the porch stood the well, which supplied water for the family's and their guests' personal consumption and for garden irrigation. Eighteenth-century Chesapeake wells were often walled with stone and sometimes were dug to a depth of

35 feet or more, so that there would always be 4 to 5 feet of good water standing in them.[20] The water was retrieved using bucket and pulley, and Faris used the ancient irrigation technique of regularly flooding carefully constructed dirt channels that ran throughout his garden, which he called "water tables."[21]

One of these irrigation paths led past an arbor. Faris planted flowering beans "round the Arber," which probably had an open-work roof to support ornamental flowering vines and defined a focal point in the garden. It may have enclosed a space for a simple bench or a more elaborate garden seat, although Faris did not write of such a seat. During the early 1790s, garden seats were being advertised for sale in nearby Baltimore, "made and painted to particular directions."[22]

For a while, a dovecote sat near Faris's arbor. In the Chesapeake, dovecotes were also called Culver-houses, and until 1798 Faris's grounds boasted just such a nesting place. But in March 1798 he noted in his diary, "the Pigeon House Blew Down, it was Built in the year 1777."[23] Faris's Culver-house was constructed as a matter of economic convenience rather than strictly as a garden ornament. He raised pigeons for consumption by his family and the patrons of his tavern. Unlike other domestic fowl, pigeons needed no confinement, because they were home-loving birds, seldom straying far from their dovecotes. Faris's pigeon house was constructed of wood and mounted on wooden posts, although more complicated colonial dovecotes existed, like the circular brick and stucco dovecote (reminiscent of the early Roman columbaria) at Tryon Palace in North Carolina. One English visitor wrote of the less elegant dovecotes he observed in the Chesapeake at the end of the century, "There are some pigeons, chiefly in boxes, by the sides of houses."[24] After pigeon consumption was no longer an essential element of the Chesapeake economy, dovecotes survived largely as garden embellishments, providing the gardener and his guests both visual and aural pleasure. (See Plate 17.)

Pidgeon

One traditional garden component on Faris's grounds was the result of a gift he received in the spring of 1793, when a neighbor "Made Me a pressent of Hive of Bees." By the next winter, Faris had built a shelter for the hive, putting "the frame of the bee house together." Faris's bee house was a painted pine box that may have been self-contained or may have served as a shelter for the more traditional but perishable straw skep; because only two years after building the wooden

box, Faris "drove the Bees out of
the Old Hive into a nother and
took the honey, the Hive was
Rotten and Ready to tumble to
peaces." But a visitor to Maryland during the same period
noted, "Honey-bees are kept in
America with equal success as in
England . . . I never saw a hive
made of straw."[25] Bees had long
been garden residents and were
considered decorative as well as
practical. In 1618 William Lawson
wrote in *New Orchard and Garden,*

"There remaineth one necessary thing . . . which in mine Opinion makes as much
for Ornament, as either flowers, or forme, or cleanness . . . which is Bees, well
ordered." The ever-practical craftsman, Faris knew that bees served him well as
both pollinators of plants and producers of honey and were worth the trouble of
keeping them "well ordered."

A few years earlier Faris's garden had sported another traditional functional
garden component, a rabbit warren. Even though the rabbits graced his family's
table for many years, he in time dispensed with keeping them. In 1792 he noted
his intention to remove "the fence from the Rabbit yard and . . . take up the
Bricks." The rabbits' place on the grounds was eventually usurped by an additional vegetable plot.[26] Faris may have found raising rabbits to be less cost effec-

tive than raising produce, for one English visitor to the Chesapeake was skeptical of the possible success of raising rabbits for food or profit in Chesapeake gardens: "Mr. Smith had got some imported rabbits . . . from England, with an intention to make a warren; but this will not answer in any part to America that I have seen. . . . First, there is no sod to make banks; therefore the fence must be all paled to keep them in, which is an enormous expense. Secondly . . . the winter is so severe they would not pay for the food they would eat."[27]

The most surprising item in the practical craftsman's garden was a purely ornamental embellishment, a statue. Classical statues reminiscent of gardens in the Italian Renaissance dotted the grounds of wealthier Marylanders during the period. One of the Revolutionary War heroes to whom Faris had dedicated a tulip was Colonel John Eager Howard, whose Baltimore home was renowned for the statues that graced its gardens.[28]

Milking the Cow

Faris's grounds contained a practical structure he politely referred to as the "temple" in his garden. While some Chesapeake gardens may have had miniature versions of temples built on their pleasure grounds, Faris's temple was his "necessary," which he also nicknamed the "little house" and around which he consistently planted flowers in rectangular beds carefully bordered by boxwood. As concern for basic survival in the British American colonies decreased, concern for propriety increased. One Maryland acquaintance of Faris wrote, "Many in-

Gardner and Barrow

stances there are of a scandalous neglect of decency, even in opulent farmers, in their not building a single necessary. . . . such ought to be provided wherever there is habitation, be the family many or few, rich or poor."[29] Early Americans determined the placement of the privy by some compromise between convenience and the senses. A German traveler touring the Chesapeake in 1783 noted that behind most town dwellings in America "is a little court or garden, where usually are the necessaries, and so this often evil-smelling convenience of our European houses is missed here, but space and better arrangement are gained."[30] A strictly utilitarian shed, 16 by 20 feet, sat near the family privy.[31] In it Faris stored his simple gardening tools, which included a spade, trowel, hoe, and rake.

The outbuildings of town homes in the eighteenth-century Chesapeake often bordered and helped define the garden. Stables were usually the farthest removed outbuilding from the house. A red-and-white milk cow was the only permanent resident of Faris's stables during the 1790s, but they served as temporary home to the horses of guests at the inn. Several of Faris's neighbors had "chaise houses" separate from their horse stables, to contain their carriages.[32] Faris planted most of his kitchen garden beds and some flowers near his stables, as contemporary Chesapeake garden writers advised. Dung was the fertilizer of choice in the eighteenth century. Faris consistently used dung from his own stables and employed neighborhood haulers to bring extra cartloads of "tan" to his garden throughout the growing season. Farmers in the Chesapeake countryside sometimes dug fenced dung pits near their "cow houses" to systematically collect future garden fertilizer.[33]

Also producing dung were the pigs Faris raised in a hog pen on the rear of his grounds, near his peach trees. Faris cooked his peach-flavored pork as it was killed and also smoked it. From the beginning of the eighteenth century, travelers throughout the Chesapeake reported, colonists in the region intentionally fed peaches to their pigs to produce a sweeter-flavored meat. On October 3, 1777, British soldier Thomas Hughes reported that "at this time fruit is in such plenty that their hogs are fed on apples, peaches, and chestnuts."[34] One of the gentlemen who bought flower bulbs from William Faris, Captain John O'Donnell (1749–1805), settled in Baltimore, naming his country seat after his favorite port of call, Canton. An account of Canton given by a visitor noted that O'Donnell had planted orchards of red peaches on his 2500-acre estate in hopes of manufacturing brandy for trade but had met with limited financial success, "For although Mr. O'Donnell's orchard had come to bear in great perfection and he had stills and the other necessary apparatus, the profit proved so small that he suffered the whole to go to waste and his pigs to consume the product."[35]

In addition to pigs and peaches, the rear of Faris's lot also contained his smokehouse, which was surrounded by plum, pear, mulberry, cherry, almond, and apple trees. Grape vines grew in one corner, near the vegetable beds. Currant and gooseberry bushes dotted the back lot as well. Faris used his one-story brick smokehouse (12 by 10 feet) to smoke both pigs and fish. Smoking dehydrated the meat, added a desirable taste of wood smoke to the final product, and allowed the fish and pork to be kept longer. One traveler through Maryland in the 1790s wrote, "The greater number of people in America live on salt fish and smoked bacon: and the reason why they smoke their bacon and fish, is, that there are many sorts of reptiles that would absolutely destroy it, were it not for the smoke."[36]

DUNG CART

Even though economy of space demanded that Faris use his grounds in a practical way, he took pride in decorating special focal points in his garden with several kinds of moveable plant containers. His favorites were earthenware pots. He regularly refilled all of his plant containers with "new dirt." Faris singled out the plants he considered rare to put in pots around his grounds, annually potting Jerusalem cherry trees, ice plants, egg plants, and sensitive plants, as did Thomas Jefferson. Faris also regularly displayed mignonette, tuberose, asters, anemones, polyanthus, rosemary, hyacinths, impatiens, chrysanthemums, and his favorite tulips in containers.

He used the pots to store his fragile plants away from the Annapolis winters, dutifully recording in his diary each year, "I moved the Potts into the seller for the Winter." Sometimes he euphemistically referred to his cellar as "the greenhouse." Faris had no greenhouse; but his Annapolis neighbor Dr. Upton Scott (1724–1814)

did, and the two men exchanged hundreds of plants. A contemporary wrote of Scott, "He is fond of botany and has a number of rare plants and shrubs in his greenhouse and garden."[37]

Faris's gardens also sported large flower-filled wooden half-barrels, which dotted the grounds. He called these unpainted containers "casks" and artfully planted them with ice plants, egg plants, Jerusalem cherries, tulips, wallflowers, India pinks, and tuberose.[38] Faris made no attempt to move his casks indoors for the winter season but did regularly change the earth in the containers. It is likely that these casks were old shipping barrels from the Annapolis docks.

The more mundane plants Faris put in simple rectangular wooden boxes.[39] These were strictly utilitarian containers, not the more ornamental wooden boxes holding orange and lemon trees that could be found in the greenhouses of larger Chesapeake plantations of the period (see Plates 20 and 21). In these boxes Faris also experimented with growing new varieties of plants, from cabbages to tulips. In his experiments, Faris grafted and selectively cross-pollinated plants. Gardening in the eighteenth-century Chesapeake allowed every man to become his own scientist, as the Italian Renaissance model promoted.

This artisan, innkeeper, and gardener was keenly aware of the changes in nature's seasons that intimately affected the success or failure of his gardening efforts. He even noted in his diary when the martins returned to Annapolis.[40] Like Washington's and Jefferson's records, his diary recorded his observations of the weather, and he consistently referred to his notes when new plants broke through the ground, or when they bloomed—or when they failed—in order to compare present efforts with previous attempts.

Like his wealthier gardening colleagues, William Faris used his garden to project his abstract ideas into nature. He and his neighbors used their gardens to understand the order of nature and to subject it to their own order in terms of design, plantings, and processes.

Gardens of the Gentry

House and garden tours are not a twentieth-century phenomenon. Eighteenth-century Chesapeake gentry and artisans alike enjoyed viewing gardens on their journeys and in their local neighborhoods and believed that one could tell a lot about people by the gardens they kept. Serious gardeners even recorded their observations. William Faris, strolling the streets of Annapolis, jotted notes in his journal about his neighbors' gardens.

John Adams was an inveterate garden watcher and often judged the status of his contemporaries by what he saw. He took note of Baltimore gardens in the winter of 1777, when the Continental Congress met there. At William Lux's 1750s country seat, Chatsworth, he noted, "the large garden enclosed in lime and before the yard two fine rows of large cherry trees which lead out to the public road. There is a fine prospect about it. Mr. Lux lives like a prince." The princely grounds, which included a 164-by-234-foot terraced garden, were later sold to become a commercial garden and renamed Gray's Garden.[1]

Lux chose to protect his investment by building a stone wall directly connecting Chatsworth's central-axis, symmetrical garden to the house, a feature it shared

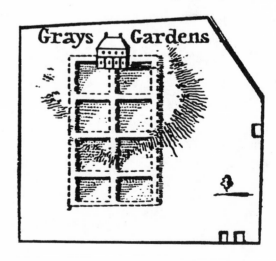

John Adams, in Baltimore with the Continental Congress, admired William Lux's Chatsworth grounds, which later became a public pleasure retreat named Gray's Garden.

The garden at the Lux estate was probably measured by use of an eighteenth-century device called a Gunter's chain (*upper left*), which was 66 feet long. The unit of measurement called a perch was one-quarter of a chain, or 16 feet. Lux's garden would have measured about 2½ chains by 3½ chains. Woodcut from Thomas Hill, *Gardener's Labyrinth*, London, 1564.

with several of the earliest Chesapeake gardens. Holly Hill, Maryland's oldest surviving seventeenth-century brick house, had a geometrically balanced walled garden directly adjoining the L-shaped building.[2] Bacon's Castle, Virginia's earliest seventeenth-century brick dwelling, was an exception. It's rectangular garden was only partially walled and not connected to the main house. The Bacon's Castle garden was set off to the side and was not a Palladian progression of the geometric lines of the dwelling.[3]

Inside its brick wall, William Lux defined Chatsworth's grounds by creating eight equal-sized rectangles, or "oblongs" as English garden authority Philip Miller called them. Miller, whose work *The Gardeners Dictionary* was widely read in the colonial Chesapeake, recommended central-axis gardens with matching squares on either side of a gravel walk leading out from a door at the center of the house. Most colonial gardeners adopted Miller's advice and built their garden beds twice as long as broad. By mid-century, main-axis symmetry dominated most mid-Atlantic gardens.

William Lux planned Chatsworth's walled garden before the Revolution, in the still wild Baltimore countryside, and he probably felt safer with the control a wall afforded him. His garden was reminiscent of medieval European walled gardens, which closed out interlopers and in which humans molded nature to their own uses. In later European walled gardens, the owners often toiled to perfect the Neoplatonic ideal of producing perfect examples of flowers and rare plants.

The oldest surviving seventeenth-century house in Maryland, Holly Hill in Anne Arundel
County, featured a walled and balanced garden adjoining the house, effectively expand-
ing interior space. To the designers of this estate, perfect symmetry seemed unnecessary.
From a painting, ca. 1730, Maryland Historical Trust file AA-268.

There visitors could admire specimens of imported exotic as well as native plants
either in pots or planted directly in the soil. And so it was in the colonies. Men
and women alike became plant hunters in seventeenth- and eighteenth-century
North America. They excitedly exchanged plants and sent new species back to
England and Europe for study. William Byrd, in his diary entry for April 10, 1720,
wrote of entertaining guests at Westover in Virginia: "After dinner we walked in
the garden and I showed them several rarities."[4]

By mid-century, city-dwelling gentry, in the quickly growing towns up and
down the Atlantic, often built brick-walled gardens as well. Charles Carroll of
Carrollton incorporated a wall into his grounds in Annapolis in 1774; his neighbor William Paca (1740–99) had enclosed his garden with a brick wall nearly a
decade earlier.[5] Brick walls were expensive and available to only a few in early
America, but they served useful purposes. They kept vegetables and fruits safe
from intrusion, and they announced that the owners were persons of means.

The Paca House garden, reconstructed in the 1970s, is unusual in that its main
walk does not lead from a center door on the garden façade, so the garden does
not sit on a central axis, relative to the house. The garden, for years buried under
a paved parking lot, was restored as a typical geometric and symmetrical garden
on the top terraces. The lowest terrace Paca designed with a lake and a summerhouse in a contrived naturalistic style. Charles Willson Peale included this lower
terrace in a portrait of William Paca, and it is the only documented space in the

garden.[6] Paca had just returned from England when he began building his garden in the 1760s, so he would have been familiar with the natural style in vogue in Britain at that moment.

Paca's house, also built in the 1760s, was not at all large compared to the homes of English gentry, but for Annapolis it was quite grand. The brick structure comprises a 48-by-44-foot two-story center section and two single-story wings—a kitchen wing of 32 by 16 feet and a corresponding office wing of the same size. Paca purchased two 198-by-198-foot lots for his house and gardens. The gardens he planned consisted of three falls, narrowing as they dropped $16\frac{1}{2}$ feet to the naturalized lowest level featuring the summerhouse and a Chinese-style bridge over a pond (see Plates 14, 15, and 16). Archaeologists found that the terrace closest to the house measured 80 feet in width, the next 55 feet, and the last 40 feet. This design allowed those viewing the two-acre garden from the house to see grounds that appeared larger than reality.[7] Using optics to create an illusion of larger houses and grounds was particularly important in colonial Chesapeake towns, where space was limited, but the need to appear important was boundless.

Chesapeake gentry considered Oriental embellishments, such as the bridge in the Paca garden, high style in this period. In 1762 Philadephia diarist Hannah Callender wrote of a local garden, "In the midst is a Chinese temple for a summer house."[8] Using Oriental designs signaled to those passing by and stopping in that the owner was a refined, genteel leader in society. Both summerhouses and temples served as social gathering sites in eighteenth-century America. Although in the Chesapeake they often sat on naturally elevated grounds, Paca placed his at the foot of his terraces.

The William Paca House garden in Annapolis was excavated and reconstructed. Adapted from Historic Annapolis materials.

Sloping falls gardens, such as Lux built in Baltimore and Paca built in Annapolis, could be found up and down the Atlantic Coast throughout the eighteenth century. Because the topography of the area allowed it, many Chesapeake gentry whose homes sat on a rise of ground terraced their gardens. Many of these falls sloped down to a body of water, and the main approach to colonial houses was often by water.

Aesthetically, terraces provided a setting for the house, a pleasing view from upper stories, and a platform for surveying the surrounding countryside. A contemporary American garden authority acknowledged the garden as a stage when he wrote "regular terraces either on natural eminences or forced ground were often introduced . . . for the sake of prospect . . . one above another, on the side of some considerable rising ground in theatrical arrangement."[9] Such designs elevated the wealthy owner above the common audience passing by or strolling through. One look at nature so well ordered and the observer could have no doubt that here lived a person destined to be in charge.

Charles Carroll of Carrollton, a signer of the Declaration of Independence, and his father were spending the decade of the 1770s worrying about the political direction of the colony and designing geometric gardens for their Annapolis home. Their gardens covered about 2¾ acres. Broadening terraces fell 24 feet from the house to Spa Creek. The garden terrace closest to the water was 50 feet wide, the next ascending terrace 40 feet wide, and the garden terrace, closest to the 45-foot-long house, measured 30 feet in width. This plan made the three-story house seem even more imposing when viewed by visitors approaching from the water, and made the water seem closer and broader when viewed from the house. The younger Carroll realized that a perception of superiority could work to his detriment in society. In a letter to a friend he wrote, "There is a mean low dirty envy which creeps thro' the ranks and cannot suffer a superiority of fortune, of merit, or of understanding. . . . my fortune will certainly make me an object of envy."[10] Nonetheless, it was an unsteady time in the colonies, and a visual representation of power and control probably would not hurt.

At the bottom of the terraces, where a walkway ran 400 feet along the water's edge, the Carrolls placed octagonal summerhouses, at each end of the walkway. "I like my pavillions," wrote the younger Carroll, "they are rather small." Between the pavilions, ladies often fished off of the walkway.[11]

Many colonials referred to the level area of a terrace as "the flat." They would plant these flats either in turf or in garden beds. The latter could include ornamental flowers as well as the useful vegetables and herbs that the Carrolls chose for their flats.

In the first half of the century, colonial gardeners sometimes created intricately designed beds of flowers. The most ambitious early gardeners attempted flower

Opticks

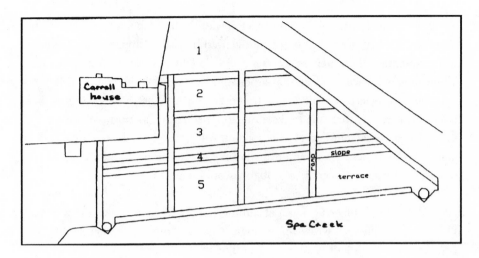

The Charles Carroll gardens in Annapolis fell in terraces from the house down to the water, where a 400-foot walkway ran along the water's edge, anchored by two octagonal pavilions. Drawing by Elizabeth Kryder-Reid, used by permission.

knots. In 1749 a house for sale advertisement touted "a garden, genteelly laid out in walks and alleys, with flower-knots, & laid round with bricks."[12] Flower knots were beds formed into curious, complicated, and fanciful shapes meant to please the eye, especially when seen from a higher elevation. Flower knot designs sometimes imitated the intricate patterns of the embroidery and cut work executed by needleworkers of the time. The length of the flower knot bed was generally about one and a third times the width. Beds separated by narrow paths were usually mirror images of each other, their patterns repeated at the ends and sides of the sections created within them.

In the second half of the century, intricate flower beds became less desirable in the American colonies; then, in the 1790s, their popularity once again soared. Flowers remained a garden favorite, but gardeners in the second half of the century tended to segregated flowers by type rather than integrating them into a complicated design. Diarist Anne Grant reported that, in the gardens she saw before the Revolution, flowers "not set in 'curious knots,' were ranged in beds, the varieties of each kind by themselves."[13] In 1789, William Hamilton instructed the gardeners at his Philadelphia estate, Woodlands, to plant "exotic bulbous roots . . . at six or eight Inches from each other . . . taking care to preserve the distinctions of the sorts."[14]

After the Revolution, a few gardeners began banishing flowers in favor of turf. Philadelphian Elizabeth Drinker wrote in her diary, "flower roots . . . were dug out of ye beds on ye south side of our Garden—as my husband intends making grass-plots and planting trees."[15] Plain grass flats often defined the terraces of the gentry. During the same period, however, professional seed merchants were enticing the growing gardening public to plant curious bulbs and roots imported from Europe and elsewhere, and this seems to be the style that caught on. By the 1790s, specimen gardens and flowers once again flourished in the Chesapeake.

Terraced gardens usually had three to five terraces, the flats of which were planted with turf, or flowers, or fruits and vegetables and the sloping fronts and

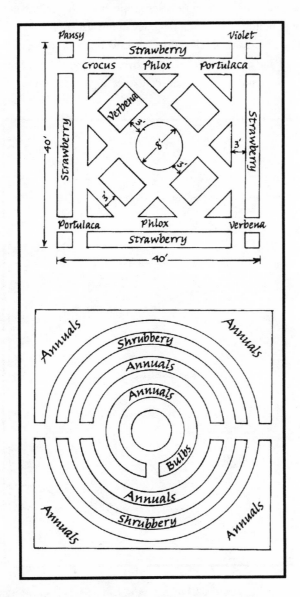

Drawing based on Lady Jean Skipwith's plans for the gardens at Preswould, in Virginia, 1790s. Skipwith Papers, Swem Library, College of William and Mary.

sides faced with turf. Garden visitors and workers moved between levels by walking up and down these grass ramps, called falls or slopes.

Falls were an early component of pleasure gardens in much of colonial America. A 1736 contract between Bostonian Thomas Hancock and his gardener reads
in part, "I . . . oblige myself . . . to layout the next Garden or flatt from the Terras
below."[16]

Falls appear very early in Virginia. Some speculate that falling gardens existed
at Green Spring, built by Governor Berkeley in the last couple decades of the seventeenth century.[17] Governor Alexander Spotswood installed terraced gardens at
the Governor's Palace in Williamsburg between 1715 and 1719. Later the governor built his private estate, the Enchanted Castle, near Germanna, where William

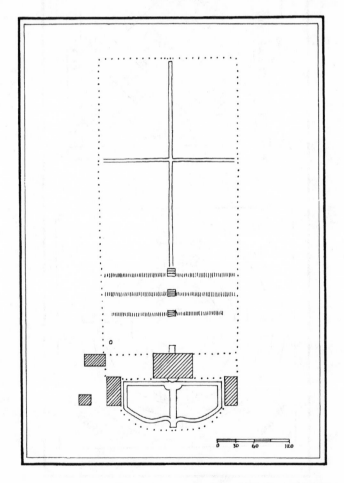

Lewis Burwell's garden at Kingsmill, built in 1730s, James City County, Virginia. Terraces
led down to an enclosed kitchen plot. Plan by William Kelso, Virginia Department of Historic Resources, from archaeological data collected by Fraser Neiman, 1975.

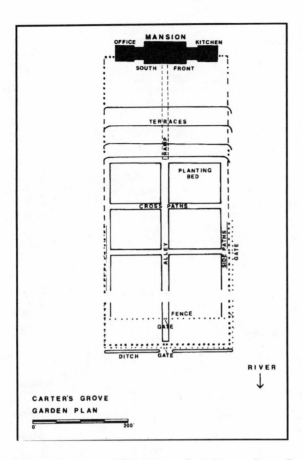

The gardens of Carter's Grove, near Williamsburg. Built in 1751, Carter Burwell's garden displayed nearly the same design as his cousin Lewis's. Site plan, Historic Williamsburg Foundation, Inc.

Byrd II came to call on September 28, 1732, and later reported, "the Garden . . . has . . . 3 Terrace Walks that fall in Slopes one below another." In the 1750s Speaker of the House of Burgesses John Robinson installed "a large falling garden enclosed with a good brick wall" at his plantation, Pleasant Hill, overlooking the Mattaponi River.[18]

It is certain that Virginian Lewis Burwell in the 1730s constructed a long rectangular garden, about 220 feet wide and extending almost 500 feet south from the house down to the James River, on his plantation, Kingsmill. Three turfed terraces led down to a large enclosed kitchen garden, which was divided into quadrants by two central walkways. Unlike the falls in Maryland gardens, Burwell's were connected by stone steps rather than grass ramps.[19] This same design was chosen by his cousin Carter Burwell in 1751 for the gardens at Carter's Grove, near Williamsburg.

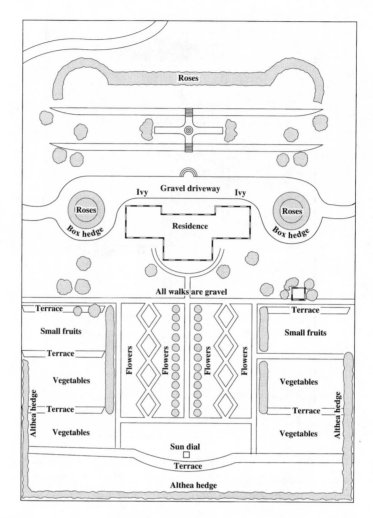

Gaymont, Caroline County, Virginia. Walkways and flowerbeds followed the downward contour of the land; vegetable beds were arranged in descending terraces. Adapted from *Historical Gardens of Virginia* (Richmond: Garden Clubs of Virginia, 1930).

Also in the 1730s, Landon Carter, a son of wealthy planter Robert "King" Carter, built similar terraces at Sabine Hall on the north side of the Rappahannock River in Virginia. (He named his estate after Horace's villa outside of Rome, Sabine Farm.) Landon Carter's riverfront garden consisted of six deep terraces spanning the width of the house. His terraces were so steep that he "almost . . . disjointed" his hip by "walking in the garden" in 1764. Steeply terraced gardens could prove deadly in fact. Charles Carroll of Annapolis died, in 1783, as a result of a fall in his garden.[20]

Terraced falls were popular among the Virginia gentry building in towns as

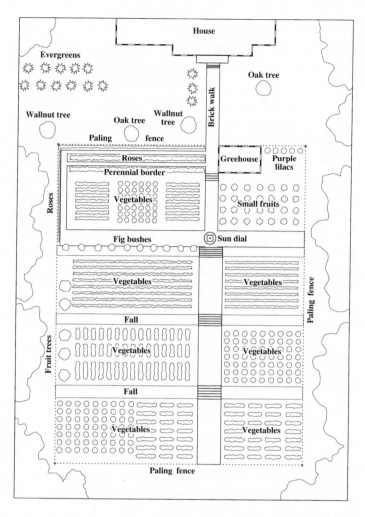

Woodberry Forest, in Madison County, Virginia, exemplified the Virginia falls garden. Adapted from *Historical Gardens of Virginia* (Richmond: Garden Clubs of Virginia, 1930).

well in the eighteenth century.[21] In the city of Richmond, Colonel Richard Adams and his son Dr. John Adams built homes with gardens falling toward the James River near old St. John's Church. Farther north, William Fitzhugh added terraced gardens to his home, Chatham, on the Rappahannock river near Fredericksburg. Nearby, Colonel Francis Thornton and his father adorned The Falls and Fall Hill with terraces to the Rappahannock. In fact, terraced falls were so admired in Fredericksburg, that in 1777 eight lots were offered for sale with the notation that four were already "well improved with a good falling garden." In 1780 another Fredericksburg newspaper advertisement touted "a good dwelling house with every conveniencey that a family can wish for . . . a falling garden."[22]

Greatly enamored of falls gardens, the gentry built them on any available natural rise, not just on riverbanks. In 1747 Colonel John Tayloe built five grassy terraces at Mount Airy, even though the Rappahannock was three miles away. Virginians kept building falls gardens well into the nineteenth century.[23]

A few English country house gardens of the eighteenth century are depicted with classical terraces on their grounds, but apparently they were the exception rather than the rule, or they were not notable enough to record. The majority of early America's terraced gardens were similar to the balanced, rectangular plan portrayed in Englishman William Lawson's early-seventeenth-century work *A New Orchard and Garden.* Chesapeake garden design was almost untouched by the excesses and ostentatious aspects of Italian Baroque and French grand manner garden styles of the time, and very little influenced by the eighteenth-century English natural grounds reaction to these formalities. Time, distance, and ideologies blunted these extremes of style.[24]

As the population grew and colonists building in the countryside felt safer from intrusion by unwelcome people and animals, homeowners began to consider the aesthetic possibilities of opening up their gardens to the surrounding landscape. Unlike their English cousins, few in British America manipulated the structure and layout of the existing natural countryside outside of the grounds immediately surrounding their homes, and few owned all the land they surveyed about them.

Up and down the Atlantic seaboard, where the topography allowed, house siting practices harked back to the defensive habit of building on the high ground. Even when the need for surveying the countryside for marauders had long passed, people continued to look for the highest situation, in part so they could remain on top in the minds of others. When choosing a homesite, gentlemen considered the vistas and views available from the pinnacle of the property. The famous architect Benjamin Henry Latrobe wrote in 1798, "When you stand upon the summit of a hill, and see an extensive country of woods and fields without interruption spread before you, you look at it with pleasure. . . . this pleasure is perhaps very much derived from a sort of consciousness of superiority of position to all the monotony below you."[25]

Visitors often used the powerful verb *command* to describe the placement of a dwelling on a site surrounded with vistas. People noted that houses on high ground were situated on an "eminence." Homage to power was due the owner. Even when the houses themselves were unfinished or left to decay, impressive sites were still admired. In June 1760 Andrew Burnaby was traveling through Annapolis and noted, "the governor's palace is not finished. . . . it is situated very finely upon an eminence, and commands a beautiful view of the town and environs." Five years later Lord Adam Gordon similarly described the still unfinished governor's house:

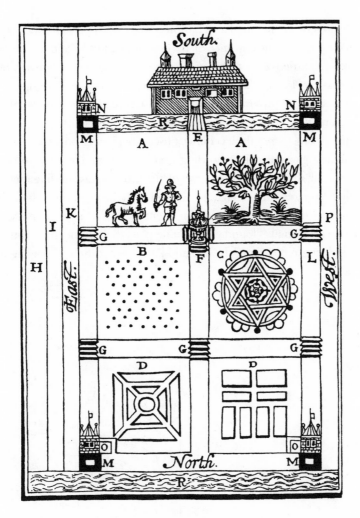

Early American terraced gardens generally resembled a plan that appeared in William Lawson's *A New Orchard and Garden* (London, 1618).

"the Situation is most Elegant . . . commanding the view of the Town, the River Severn, the Bay, and all the Creeks."[26]

In British America, gentlemen continued to construct mounts well into the eighteenth century. The gardens at George Mason's Gunston Hall in Virginia are flanked by mounts overlooking the Potomac River. British Colonial Secretary William Eddis wrote of a house in Annapolis on October 1, 1769, "The garden is not extensive, but it is disposed to the utmost advantage; the center walk is terminated by a small green mount . . . commands an extensive view of the bay and the adjacent country."[27]

One gardener who chose to build his mansion high on a hill was Charles

Ridgely (1729–90). He planned for Hampton a curved approach winding through level turfed ground to the entrance of his new house, but he laid out the majority of Hampton's gardens in a terraced design. At the mansion's garden façade, a turfed area 250 by 150 feet (alternately called the bowling green, the great terrace, or the south lawn) crowned three additional terraced falls, connected by grass ramps, looking down toward Baltimore, ten miles to the south.[28]

Bowling greens—smooth, level lawns used for playing bowls—capped many colonial falling gardens. Most bowling greens measured 100 by 200 feet, and many were sunk slightly below the level of the ground surrounding them. Sometimes called "squares" in late-eighteenth- and early-nineteenth-century America, bowling greens offered beauty and ornament as well as recreation. As early as 1666 southern colonials "found . . . a plaine place before the great round house for their bowling recreation." In the mid-Atlantic and South, playing at bowls

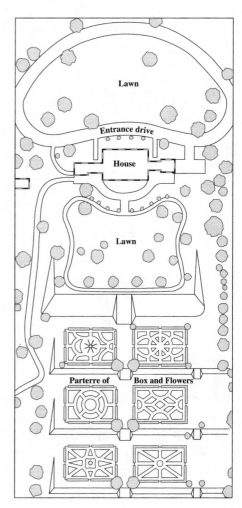

Hampton, Baltimore County, Maryland, begun in 1783. Drawn from information provided by the Hampton Historic Site.

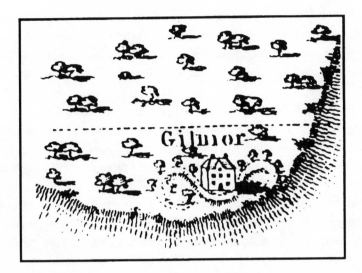

Beech Hill, Baltimore City, 1801, built by John Dorsey ca. 1770 and purchased by Robert Gilmor in 1797.

often involved wagering. William Byrd wrote in his diary on May 5, 1721, "After dinner we walked to the bowling green where I lost five shillings."[29]

At Hampton and in most Maryland falls gardens, landowners chose to build simple turfed ramps, in place of the stairways familiar in European terraced gardens as well as in a few earlier Virginia falling gardens. Ridgely planned Hampton's first garden terrace to come 18 feet below the bowling green; the second dropped a further 6 feet; and the third was 4 feet lower than the second. A grass walkway bisected the formal falling garden lengthwise from just below the bowling green to the kitchen garden at the base of their turfed terraces.[30] As we have seen from various examples, many Chesapeake gardeners planted fenced or walled kitchen gardens at the base of their ornamental terraces.

Baltimore did boast a few exceptions to the traditional ordered garden. Beech Hill, built by Colonel John Dorsey about 1770 and purchased by Scots merchant Robert Gilmor (1748–1822) in 1797, was laid out with apparently carefully planned natural grounds. There were no terraces, even though the property sat high above the bay with a spectacular view that was celebrated in paintings for decades. The land around Beech Hill contained an S-shaped driveway lined with rather evenly spaced trees but no accompanying rectangular beds or parterres were in evidence on the grounds.[31]

At the rear of Beech Hill was a white picket fence, probably enclosing a kitchen garden. This fence and the orderly trees lining the drive were the only reflection of any formality on Gilmor's grounds. The 190-acre property contained a three-story dwelling (the bottom story of stone and the other two of frame) which measured 44 by 18 feet. It also had a two-story brick milk house measuring 12 by 12 feet. Gilmor's son wrote that his father purchased the "rural retreat" because of its "extensive prospect, commanding a view over the City, Bay, river, and surrounding Country" and spent his summers there, from early June until the middle of September.[32]

A second Baltimore area garden giving a nod to the English natural grounds style was that of Harry Dorsey Gough (1745–1808) at his home, Perry Hall, far north of the city on the road to Philadelphia. The house was begun in 1773, and Gough purchased the 1000-acre property shortly thereafter, following the death of the first owner (see Plate 5). The road leading up to the home, which was located on the crest of a hill, was lined with trees as a traditional avenue. A white picket fence surrounded the house itself. The fence's elaborate wooden gate stood opposite the central door on the entrance façade, and a walkway bisecting the front yard led to the house. The fence followed an oval rather than rectangular course, but trees planted inside the courtyard fence were deliberately balanced and symmetrical in design.[33]

Of the more traditional four-bed gardens near Baltimore, fifteen included terraced falls. Typical of these was merchant John Salmon's home, which sat atop a Baltimore County hill adjoining the country seat of his partner, Robert Oliver. The partners' homes had mirror-image four-bed terraced gardens descending in the direction of the bay. Salmon's seat was a two-story home with a balcony and piazza facing south, overlooking his falling gardens. Happy with his view from both front and back, Salmon built a second piazza, on the entrance façade, facing north. Contemporaries reported that the situation was "admirable," commanding a view of the town, harbor, and river below. An observer reported of Salmon's five acres, "The garden . . . is laid off in beautiful falls: in it is an excellent cold bath and a milk house through which there runs a constant stream of water."[34]

A country seat probably more remarkable for its name than its falling gardens was Mount Deposit. This home was situated on about 260 acres acquired by David Harris (1752–1809), who built the house and gardens between 1791 and 1793. Harris was the cashier of the Office of Discount and Deposit, a leading banking institution of Baltimore, after which he apparently named his estate. The country seat was described by contemporaries as, "a large and elegant dwelling, containing ten rooms, besides kitchen and garrets, with extensive porticoes; a large barn, stables and granary, coach house, ice house and smoke house, two orchards of choice fruit and a well cultivated garden."[35] The approach to the entrance façade of Mount Deposit from the north allowed the visitor to look past the house toward Fells Point on the harbor. This side of the home was defined by a white picket fence that ran the width and twice the depth of the house and

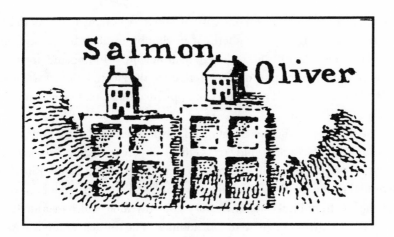

The Salmon and Oliver houses, Baltimore City, 1801.

enclosed the land immediately in front of the dwelling. Ordered plantings of trees dotted the traditional, rectangularly fenced courtyard area. (See Plate 7.) Exiting the home on the garden side, a visitor would find three terraced falls deliberately carved out of the rather steep existing slope.[36] The falls were wider than the house itself, planted with grass and shrubs but no flowers, enclosed by a white picket fence, and divided down the middle by a walk leading from the central door of the house and intersected by crosswalks. (See Plate 6.)

Northeast of Mount Deposit sat a more intricate and formal garden of four squares internally divided into circles and triangles and surrounded by a row of fruit trees on one side and a grove of trees on the other. This country seat, named Belmont, was built around 1778 by Charles François Adrien Le Parlmier d'An-nemours (1742–1809), who was the French Consul General to Maryland and Vir-ginia until 1792.[37] In that year he built "an obelisk to honour the memory of . . . Christopher Columbus . . . in a grove in one of the gardens . . . on the 3rd of August, 1792, the anniversary of the sailing of Columbus from Spain."[38]

A grove of trees in early America was a small woods or large cluster of trees, usually occurring naturally and intentionally left in the landscape or occasionally purposefully planted in the pleasure grounds around a dwelling. Often a grove consisted of large trees whose trimmed branches shaded the ground below. Groves also produced food for songbirds and served as settings for obelisks or statues meant to inspire the garden visitor. Abigail Adams wrote that Andrew and James Hamilton's Bush Hill had "a beautiful grove behind the house, through which there is a spacious gravel walk, guarded by a number of marble statues, whose genealogy I have not yet studied."[39]

Many European travelers remarked that the groves, clumps, copses, and bosques so carefully cultivated in their countries, were more easily assembled in the colonies. "In order to have a fine garden, you have nothing to do but to let the trees remain standing here and there, or in clumps, to plant bushes in front of them, and arrange the trees according to their height."[40] In England, the natural grounds movement owed part of its popularity to the fact that timber was getting scarce in the countryside. The British gentry planted their "natural grounds" with trees they needed to grow.

Two of the most ambitious traditional four-part terraced gardens near Baltimore belonged to George Grundy (1755–1825), a merchant, and Solomon Birckhead (1761–1836), a physician, whose home was called Mount Royal. Birckhead's gardens had geometric designs within its squares.[41] The garden fell in the direction of the bay from Birckhead's two-story stone dwelling, which measured 54 by 23 feet and had a two-story 31-by-18-foot addition. The 101-acre grounds at Mount Royal also contained an unusual one-story round milk house measuring 10 feet in diameter, a frame smokehouse 12 by 12 feet, two barns of two stories, one stone and one frame, both measuring 46 by 24 feet, and a two-story stone mill 51 by 41 feet. The house and its geometric gardens were built around 1792.[42]

A short distance from Mount Royal sat Bolton, which George Grundy built

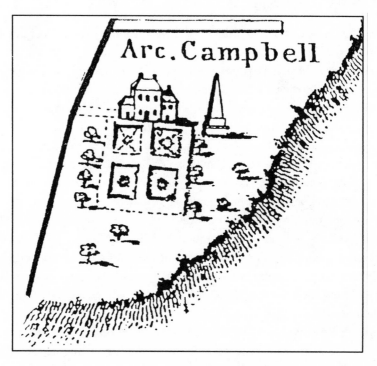

Archibald Campbell's property, purchased from the French Consul, who constructed an obelisk in honor of Christopher Columbus on the tricentennial of Columbus's voyage to America.

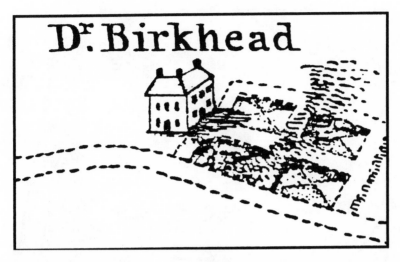

Dr. Solomon Birckhead (the name was misspelled on the map) grew medicinal plants in his falling gardens at Mount Royal, built in 1792 in Baltimore.

immediately after he acquired the 30-acre property in 1793. In addition to a 65-by-37-foot two-story brick home, he contructed a barn, two coach houses, a wash house, a smokehouse, and an icehouse, with orchards of fine peach, apple, and cherry trees. Near a spring of pure water and beside the enclosed kitchen garden area sitting at the base of his falling turfed terraces, Grundy built a comfortable two-story frame dwelling for a gardener.[43]

Grundy planned his garden to consist of three individually fenced rectangular turfed falls dropping south toward the harbor. These terraces were more than three times the width of the house, and initially the lowest rectangular terrace was planted in rows to serve as a kitchen garden. Gravel walks defined each terraced division, and a walkway ran from the house bisecting each terrace. At the turn of the century, Grundy altered the lower garden to a semicircular bed surrounded by a white picket fence that projected from the fencing enclosing the rectangular terrace just above it. In the center of this semicircle was a large flowering tree or group of shrubs surrounded by a circular walkway. Outside of this walkway and within the picket fence were rectangular beds now planted with flowers instead of vegetables and herbs.[44] (See Plate 1.)

The large flowering tree or group of shrubs in the middle of the lowest flat at Bolton may have been a living bower or arbor (see Plate 2). Colonial Americans had for decades planted trees, shrubs, flowering beans and vines for cool shade. A visitor reported in 1679, "We had nowhere seen so many vines together as we saw here, which had been planted for the purpose of shading the walks on the river side."[45] In 1787, at Grey's Gardens near Philadelphia, Manasseh Cutler reported,

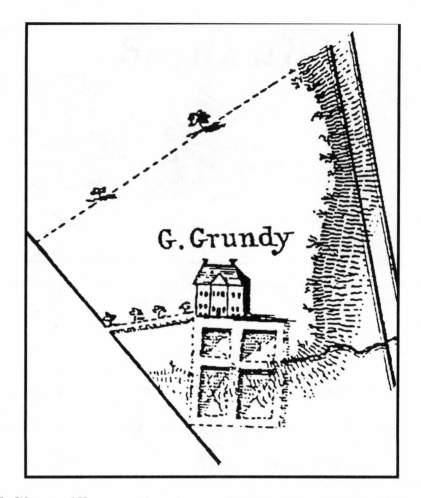

The Warner and Hanna map, drawn in 1797, reflects George Grundy's original design of Bolton. Revisions implemented a few years later made it more decorative and somewhat less utilitarian.

"At every end, side, and corner, there were summer-houses, arbors covered with vines or flowers or shady bowers encircled with trees and flowering shrubs."[46]

It is possible that the unusual large circle of shrubs or trees in the midst of Bolton's garden may have been similar to one in Salem, North Carolina, where a visitor wrote, "into the garden . . . we saw . . . a curiosity . . . extremely beautiful. It was a large summer house formed of eight cedar trees planted in a circle, the tops whilst young were chained together in the center forming a cone. The immense branches were all cut, so that there was not a leaf, the outside is beautifully trimmed perfectly even and very thick within, were seats placed around and doors or openings were cut, through the branches, it had been planted 40 years."[47]

At Bolton, George Grundy planted the approach from the north with evenly spaced tall cedars along the wooden fence bordering the property. An elaborate white picket gate opened from the public road onto Grundy's private driveway, which led to a second wooden gate, situated directly opposite the central door of the house on the north entrance façade. The house on this side was totally surrounded by a rectangle of white picket fencing. The driveway within the closest fenced area was a semicircle, with no deliberate plantings, either naturalistic or symmetrical. Bolton was a prime example of a traditional Chesapeake garden.

But Baltimore also had nontraditional gardens, such as the curious garden of five distinct rectangles owned by a French merchant John Carrière (1768–1837). This country seat was unusual not only for the number of garden beds it boasted, but also because the house, an elegant structure named Libourne, sat in the middle of the largest of the beds, where it was intimately surrounded by the glory of its garden.[48]

Several Baltimore homes sat perched above six-bed terraced gardens. One of the most sophisticated was built by Paul Charles Gabriel de Ghequiere (1754–1818), a grain merchant who designed his home and gardens upon arriving in Baltimore in 1782 from Courtney, France. Ghequiere chose an avenue of tall lombardy poplars standing like soldiers at attention along the path to the entrance façade. The intentionally ordered approach demanded respect for the owner. He built a wide gravel walk that divided the front yard. Rectangular flower beds, outlined by white picket fencing, bordered the walkway to the brick house, which

John Carrière's Libourne, Baltimore County, 1801.

Paul Charles Gabriel de Ghequiere, a grain merchant from France, lined the entrance drive to his property with poplars that stood like soldiers at attention, creating an imposing appearance.

measured 50 by 24 feet. (See Plate 13.) He planned that the doorway on the back of the house would open onto a garden of six geometric French-style parterres surrounding a fountain or water bason. Ghequiere's country seat contained 56 acres and boasted a fine view across Baltimore to the harbor.[49]

The nearby home of Colonel (later Governor) John Eager Howard (1752–1827), called Belvedere, contained a geometric garden area whose design matched Ghequiere's and boasted a spectacular view down to the harbor. Howard erected Belvedere between 1783 and 1786. The large hipped-roof mansion was noted for its magnificent gardens and for statues, which dotted the grounds just as at the papal Belvedere, where Julius II displayed his collections of statues during the Italian Renaissance.[50]

Up and down the Atlantic coast, colonial gentry embellished their grounds with statues. As early as 1718, Judge Sewall in Boston, Massachusetts complained that, "Quickly after the wind rose . . . It blew down the southernmost of my cherubim's heads at the Street Gates . . . they have stood there near thirty years."[51] In 1754, New England preacher Ezra Stiles reported that Andrew and James Hamilton's Bush Hill in Philadelphia had a "very elegant garden, in which are 7 statues in fine Italian marble curiously wrot." Then as now, opinions on the aesthetic qualities of gardens and statues were highly subjective. Of the same garden, four years later, a less enthusiastic traveler noted that it did not "contain anything that was curious . . . a few very ordinary statues."[52] Also in Philadelphia, at William Peters' garden, Belmont, a visitor in 1762 recounted: "In the middle stands a statue of Apollo. In the garden are statues of Diana, Fame and Mercury.[53] Roman gods also showed up in John Custis's garden in Virginia, where in 1772 there were stat-

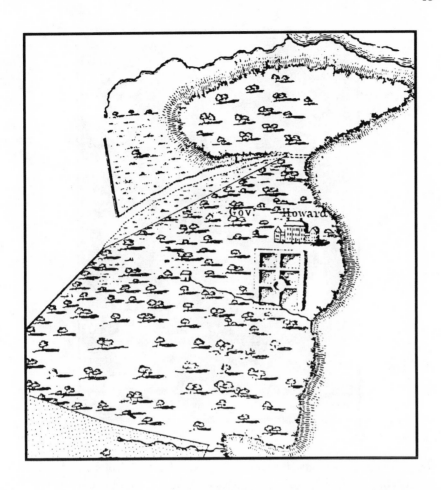

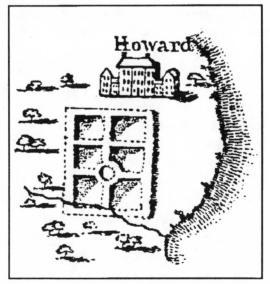

In Baltimore, John Eager Howard's Belvedere dominated high ground directly north of the town, the Jones Falls flowing below and to the west. Howard's Woods, where citizens freely strolled, gave the house an impressive setting.

ues of Venus, Apollo, and Bacchus.[54] And the garden of Colonel John Tayloe's Mount Airy had by 1774 acquired "four large beautiful Marble Statues."[55]

Some gardeners set their statues in the midst of a natural open green leading to the house, or in pairs on either side of the entrance gate. Others chose the seventeenth-century style of placing statues in the middle of a turfed garden square. In 1791 a visitor to the garden of Joseph Barrell, a Boston merchant reported seeing, "a young grove . . . in the middle of which is a pond, decorated with four ships at anchor, and a marble figure in the centre. . . . The Squares are decorated with Marble figures as large as life."[56]

Garden statues produced by American and European artists were widely available by the last decade of the eighteenth century. In Philadelphia, a 1796 ad read: "To be sold . . . Six elegant carved figures, the manufacture of an artist in this country, and made from materials of clay dug near the city, they are used for ornaments for gardens . . . they are well burned and will stand any weather without being injured . . . they represent Mars, and Minerva, Paris and Helen, A Male and Female Gardner."[57] Even non-gentry, like Annapolis craftsman William Faris, might have a statue in their gardens.

By the 1790s, Chesapeake gentry were coming up with ingenious places for their statues. If statues were meant to stand in groves of lofty trees, then why not be practical and put them in orchards, where the trees have some utility as well. And that's just what Margaret Baker Briscoe (1745–1814) and her husband Gerrard Briscoe (1732–1801) did in their orchard at Clover Dale in Frederick County, Maryland. In a 1799 portrait, Charles Peale Polk depicts Mrs. Briscoe proudly seated before a view of her orchard, in which each long row of trees is guarded by a full-sized statue perched on a marble pedestal.[58]

Statues from Europe poured into the American market. European artists were even making likenesses of American heros for export. A New York newspaper reported on a local public garden that had "lately imported from Europe . . . nineteen statues . . . Socrates, Cicero, Cleopatra, Shakespeare, Milton . . . the illustrious and immortal Washington . . . and miscellaneous figures from Greek mythology." By 1805, Vauxhall Gardens in New York City had, "procured from Europe a choice selection of Statues and Busts, . . . Washington, Cicero, Ajax, Antonious (in two poses, Hannibal, the Belvidere Apollo (in four sizes), Venus, Hebe (in two poses), Hamilton, Demostenes, Plenty, Hercules, Time, Ceres, Security, Modesty, Addison, Cleopatra (in two poses), Niobe, Pompey (in two poses), Pope, The Medici Apollo, and Thalia."[59]

Governor Howard's garden in Baltimore was remarkable for more than its statuary. Howard planned that the garden façade of Belvedere would contain a large porch, running the full length of the center block, which had a projecting two-story bay in the center. Howard enjoyed entertaining visitors for breakfast on this porch, overlooking the terraced formal gardens and the natural area of shrubs and trees that fell south toward the harbor. The entrance side of the house, on the north, faced a totally natural setting.[60]

A French visitor to Baltimore remarked on "a hill owned by Colonel Howard that dominates the town to the north. The mansion and its dependencies occupy the forward part while a park embellishes the rear. The elevated situation, the mass of trees, an appearance that evokes despite itself European ideas."[61] Apparently Belvedere owed more to its natural surroundings than it did to its large and elaborate gardens.

Garden historian Rosemary Verey has written that eighteenth-century American gardens may have retained their formality because "in England the countryside had already been tamed by years of husbandry, while in America each new plantation was surrounded by wild, untamed land, to be kept at bay, not emulated." Others, such as Elizabeth Pryor, have speculated that the alluring beauty of the natural landscape surrounding the Chesapeake Bay may help explain why gardeners were not seduced by the naturalistic style sweeping England.[62] The Chesapeake woods, continuously cleared of underbrush by Indian fires, already resembled the "improved" landscape in the watercolors of English landscape architect Humphry Repton. In fact, another visitor to Howard's home wrote: "its grounds formed a beautiful slant toward the Chesapeake. From the taste with which they were laid out, it would seem that America is already possessed of a . . . Repton. The spot thus indebted to Nature and judiciously embellished was as enchanting within its own proper limits as in the fine view which extended far beyond them. The foreground possessed luxurious shrubberies and sloping lawns; the distance, the line of the Patapsco and the country bordering on the Chesapeake." Another visitor to Belvedere claimed to "rejoice in the vistas and the sensations they inspire."[63]

Among the Howards' oldest neighbors was an Irish physician, Dr. Henry Stevenson (1721–1814), who had one of the earliest terraced gardens in the Baltimore area. His grounds displayed a flat four-bed garden on the north side of his home, called Parnassus, which Stevenson started constructing in 1763 and completed in 1769. On the south side of the house, facing the harbor, he built a bowling green and five grass terraces. A roadway wide enough for a horse-drawn wagon bisected the grass terraces up to the bowling green, and Stevenson planted double rows of trees across the width of the house, creating alleys along the outer edge of the fences that lined the terraces up to the house.[64] (See Plate 11.)

Two eight-bed gardens flourished in Baltimore County at the end of the eighteenth century. William Lux's walled garden at Chatsworth was the earlier, built in the 1750s. The second eight-bed garden graced the home of William Gibson (1753–1832), the Baltimore County Clerk, who built Rose Hill in the early 1790s.

Dr. Henry Stevenson's Parnassus, built 1763–69. Kitchen gardens lay to the north, a bowling green and grass terraces to the south.

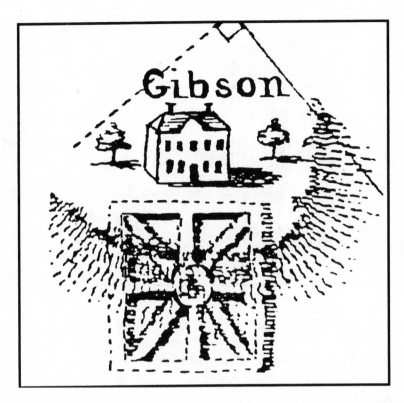

Rose Hill in Baltimore City, the home of William Gibson, stood on high ground that fell off in eight triangular beds

Rose Hill's formal gardens were falling terraces consisting of four squares divided diagonally by walkways, creating eight triangles. A fountain or water bason sat in the middle of the formal gardens, and the walkways radiated from it like rays.[65]

Near the end of the century, John Donnell (1752–1827) began the construction of his Baltimore gardens and pleasure grounds called Willow Brook. Like many other Baltimore estates, the entrance to the 26-acre country seat followed a traditional design, planted with avenues of trees and outlined with rectangular white picket fences. But adorning the entrance façade of the house were four statues sitting on classical pedestals.[66] Contemporary observers reported that the property was "divided and laid off into grass lots, orchards, garden . . . with the greatest variety of the choicest fruit trees, shrubs, flowers . . . collected from the best nurseries in America and from Europe . . . with vegetables of all kinds. . . . In the garden is a neat wooden house . . . a gardener's house, wash house, spring house, stable and carriage house, a fish pond well stocked with fish, and an elegant bath with two dressing rooms, bath and spring house."[67]

Another Baltimore country seat adorned with statues was built by Charles

Carroll the Barrister (1723–83) in the early 1760s. Mount Clare stood just a mile from the Patapsco River. Its entrance façade was surrounded by a semicircular white picket fence extending from the dependencies to a spot directly in front of the main doorway. Statues of lions sat on pedestals on either side of the walkway leading to the central door. The terraced garden façade of Mount Clare was popular with visitors.[68]

In 1770, Virginian Mary Ambler visited Mount Clare and recorded that she "took a great deal of Pleasure in looking at the bowling Green & also at the . . . very large Falling Garden there is a Green House with a good many Orange & Lemon Trees just ready to bear. . . . the House . . . stands upon a very High Hill & has a fine view of Petapsico River You step out of the Door into the Bowlg Green from which the Garden Falls & when You stand on the Top of it there is such a Uniformity of Each side as the whole Plantn seems to be laid out like a Garden there is also a Handsome Court Yard on the other side of the House"[69] (see Plate 12).

At Mount Clare, as in many eighteenth-century households, the wife supervised the greenhouse activities, while the husband oversaw the design and the more practical aspects of the grounds and gardens. Margaret Tilghman Carroll was renowned for her orange and lemon trees. Wishing to make a present of some of her specimens to General George Washington, including one tree that produced both lemons and oranges, on October 29, 1789, she sent by boat twenty pots of lemon and orange trees and five boxes of assorted other greenhouse plants to Washington at the harbor at Alexandria.[70]

Garden watcher John Adams, in Baltimore for a session of the Continental

Congress in February of 1777, spoke highly of Mount Clare. "There is a most beautiful walk from the house down to the water; there is a descent not far from the house; you have a fine garden then you descend a few steps and have another fine garden; you go down a few more and have another."[71]

The eighteenth-century pleasure gardens growing on the hillsides of the Chesapeake Bay were strikingly similar to the pleasure gardens that dotted the hills of Rome during an earlier republican era. Even town gardens of the middling classes harked back to classical precedents. The garden was the gentleman's stage and a device with which to help define his position in the emerging republic. Order, control, and regularity dominated garden designs as landowners structured their external environments to project a positive image of themselves to passers-by. The nineteenth century would see gardens grow less and less formal, and in the 1840s Andrew Jackson Downing would vigorously promote an American natural grounds movement.

3

The Republican Garden

In 1789, Jedidiah Morse, noted clergyman and geographer, wrote of one country seat: "Its fine situation . . . the arrangement and variety of forest-trees—the gardens—the artificial fish-ponds . . . discover a refined and judicious taste. Ornament and utility are happily united. It is, indeed, a seat worthy of a Republican Patriot."[1] In the early Republic, gardeners strove for a balance of useful plants and trees and genteel design. On both town and country plots, most Chesapeake gentry, merchants, and artisans planned gardens that were both practical and ornamental. Charles Carroll of Annapolis was one of the wealthiest men in the colonies, but he planted the beds of his terraced gardens with an eye toward practicality. Orderly squares filled with vegetables surrounded by low privet hedges decorated the flats of Carroll's falling terraces. Painter Charles Willson Peale reported, "the Garden contains a variety of excellent fruit, and the flats are a kitchen garden."[2] Included among the vegetables Carroll grew in the flats of his terraces were early York, battersia, red, and green savoy cabbage; white, purple, and green broccoli; cauliflower; solid and upright celery; green and white endive; green and brown Dutch lettuce; several sorts of beans and peas; round and prickly spinach; prickly, early, and long prickly cucumber; white and silver corn; spanish onion; salmon radish; mustard; cresses; and marrow.[3]

By 1798 more than seventy combination pleasure and kitchen gardens were thriving in and near Baltimore. At the end of the eighteenth century, Baltimore's well-to-do often maintained a country house in addition to their city home, to escape the diseases and oppressive heat that seized the port town in the summer months. These country seats were usually only a mile or two from town, allowing a businessman to travel to his office in town as conveniently as possible.

Generally, Maryland country seat gardeners shared John Adams' negative attitude toward the excesses of the natural grounds movement of the English. Until recently most garden historians believed that by the end of the century, few formal gardens with their traditional geometric bed designs remained in Britain.[4] In the eighteenth-century, Chesapeake gentry who gardened were well aware of the new English style and sometimes added serpentine entry roads and paths that meandered through the wooded edges of their grounds, but they overwhelmingly designed their gardens as traditional squares.

Thanks to the comprehensive mapping of "Warner and Hanna's Plan of the City and Environs of Baltimore," we are able to know the layouts of most of Baltimore's larger gardens. Of the seventy geometric plots dotting the Baltimore hillsides, gardens with four beds, probably planted with fruits and vegetables, were by far the most numerous, more than thirty-five in number and varied in design. Gardens divided into quadrants but not terraced and with few other embellishments appeared at thirteen homes. At least one of these kitchen gardens had a stone wall surrounding its four beds. One unusual example was at the house of painter and glazier James Walker, whose four garden beds were not arranged in the typical square. Three long rectangular beds were lined up side by side and a fourth was set perpendicular to them and in front of the house.[5]

Four Baltimore country seats were possessed of a single large rectangular garden plot, which probably served as the traditional kitchen garden. One of them was the home of tailor William Hawkins, which sat on the busy road that ran to Frederick. Growing fruits and vegetables was considered a necessity at country seats, since many of Baltimore's well-to-do moved to these residences during almost the entire growing season. Nine Baltimore homes had gardens of twin rectangular beds, and two of these were terraced, falling toward the harbor. Two of the remaining homes with double-bed gardens had accompanying avenues of trees leading to the house, like that of court crier William Bigger, or lining the road before it. Only two Baltimore gardens consisted of three matched rectangular beds and neither was terraced.[6]

Even the terracing of gardens itself served both aesthetic and practical purposes in the colonial Chesapeake. On uneven hillsides, terraces created flat areas for planting and helped control erosion. In 1772, Charles Carroll of Annapolis wrote to his son, who was improving his gardens on a bank of Spa Creek, "If you wish to make a continental slope from ye Gate to ye wash house, I apprehend the Quantity of Water in great Rains going ye way may prove inconvenient." He was still fretting about those garden slopes in 1775 as he wrote his son again: "Examine the Gardiner strictly as to . . . Whether he is an expert at levelling, making grass plots & Bowling Greens, Slopes, & turfing them well."[7] The elder Carroll was well aware that falls were functional devices which could divert water drainage and reduce soil erosion. Pragmatic Chesapeake landowners often constructed their terraces when the dwelling house was newly built, so that the earth, clay, and rubbish that came out of cellars and foundations could be used to shape the falls. The same practicality sometimes prompted Chesapeake landowners to build mounds. The products of cellar digging, heaped up into a mound, could be used as the base for another structure, such as a summerhouse or detached library, as well as

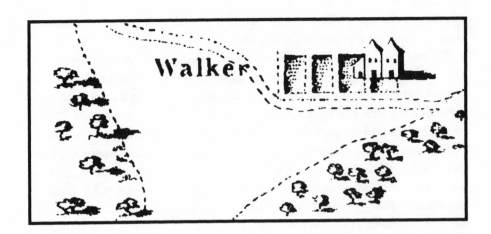

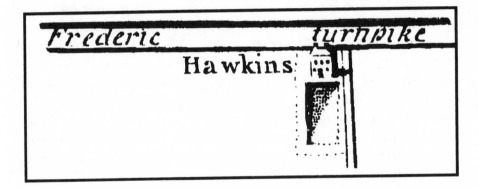

an elevated site for surveying the surrounding landscape or just a spot for catching cool air in the summer.

Most of the gardens dotting Baltimore's hillsides contained no mounds, however, and were not terraced. Typical of the simple, flat four-bed garden was the one envisioned by Colonel Nicholas Rogers (1753–1822). As Rogers planned a new home in the late 1700s, he designed the four-part garden at the back of his house with an eye to both utility and decoration. Order and symmetry dominated Rogers' plan. The garden beds were each 80 by 62 feet, separated and surrounded by garden walks 10 feet wide. An additional 10 feet of land bordered the garden walks. Down the exact center of these verges, Rogers planted fruit trees at 20-, 22-, and 25-foot intervals. Fruit trees on the property included 25 apple, 20 peach, 10 pear, 6 quince, 5 cherry, and 2 plum. In the middle of each rectangular garden bed sat one peach tree, and an additional peach tree was planted in each corner.[8] This style of placing trees was typical in colonial gardens and was called a quincunx.

Charles Carroll of Annapolis advised his son to plant his beds in quincunx. In

1777 Carroll gave his son privet rather than boxwood to outline his new garden beds and advised him to keep the privet trimmed to a small size, "not to Exceed 12 inches in Width." Thomas Jefferson's brother-in-law Henry Skipwith advised a young gardener in 1813 to consult Virgil to learn about a "quincunx, which is nothing more than a square with a tree at each corner and one in the center and thus continued throughout the orchard." Adopting classical forms was common in early American gardens. Jefferson himself wrote, "I should prefer the adoption of some one of the models of antiquity which have had the approbation of thousands of years."[9]

Of the two garden plots closest to Rogers' house, one contained a combination of vegetable and melon plants, and the second was dedicated solely to fruit.

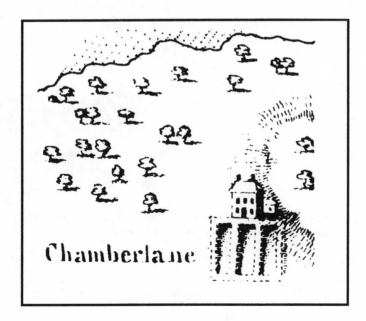

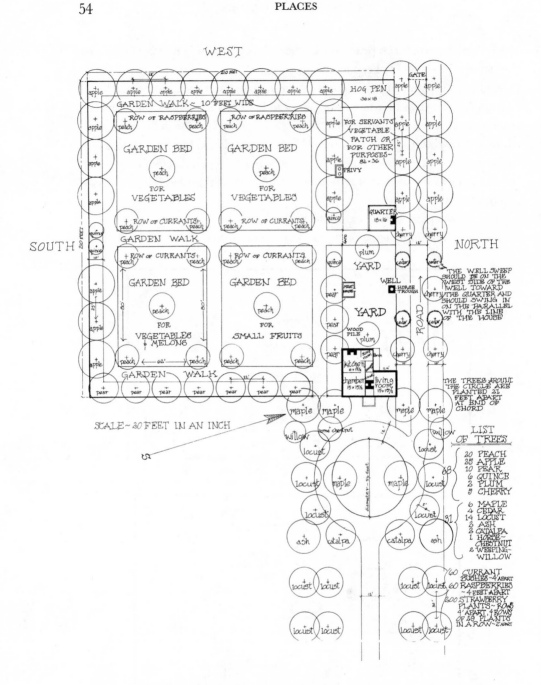

Garden plan for Druid Hill, estate of Colonel Nicholas Rogers, Baltimore, 1801. From the orginal plat, Maryland Historical Society; drawing by Susan Wirth.

The two quincunx plots farther from the house were devoted largely to vegetables. The four beds were bordered by 60 currant bushes placed 4 feet apart in single rows along the center garden walk. In the rear beds, 60 raspberry bushes 4 feet apart in single rows ran along the rear garden walk. Rogers also planted 200 strawberry plants in rows 2 feet apart in the fruit garden.[10] (Planting many fruit-bearing shrubs and trees in the garden occasionally had its downside. In 1720 William Byrd ate "so many plums" on a walk one evening that he "could not sleep."[11])

Rogers' house was 36 feet wide and sat in the extreme northeast corner of the 210-foot square that enclosed the four gardens plus the three utility areas that ran to the back of the property directly behind the house itself. A 36-by-40-foot yard containing the woodpile sat immediately behind the house. The meat house and the well signaled the beginning of a second yard, also measuring 36 by 40 feet, that continued without interruption from the first. Across the garden walk from this second yard, Rogers dedicated a plot measuring 36 feet by 82 feet to his servants. Within this area was an 18-by-16-foot slave quarter. Rogers' slaves used the remainder of this long rectangular area as their kitchen garden. The kitchen garden for slaves was usually referred to as a "huck patch" during the period. At the end of the slaves' vegetable patch, a 36-by-18-foot hog pen abutted the rear of the property. A two-seat privy stood along the outside verge of fruit trees that separated the formal garden plots from the servants' garden.[12]

Rogers planned a utility road to run just outside the 210-by-210-foot main grounds on the north side. The road was 12 feet wide with 10-foot verges on either side also planted with fruit trees at 20-, 22-, and 25-foot intervals. Just opposite the well house on the utility road side of the main grounds, Rogers built his horse trough, handy to both the well and the road.[13]

Rogers' private road approaching the entrance façade of the east-facing house was also 12 feet wide with 12-foot grass verges lining each side. About 30 feet directly in front of the entrance sat a circle of grass 56 feet in diameter. The entrance road encircled the grassed area. Two maples were the only plantings within the circle, but matching double rows of trees shaded the approach to the house. According to his written list, Rogers' ornamental trees included 6 maple, 4 cedar, 2 weeping willow, 1 horse chestnut, 2 catalpa, 2 ash, and 14 locust.[14]

Fruit trees were not the only practical yet attractive trees used in planting. Nut trees and sugar maples offered important benefits. As agricultural enthusiast John Beale Bordley observed, "The maple is a handsome clean tree. A grove of them, two or three acres, would give comfortable shady walks, and sugar for family use."[15]

Along his utility road, Rogers planted practical rows of cherry trees, like those that William Lux had planted nearly forty years earlier to lead the way to his Baltimore home, Chatsworth. Alleys of fruit-bearing trees were common in colonial gardens throughout the eighteenth century. Many Chesapeake plantations and country seats included avenues or rows of trees well into the last decade of the eighteenth century, long after such linear plantings were being shunned in England.

Homeowners planned avenues approaching their plantation houses as wide, straight roadways lined with single or double rows of trees and usually cutting symmetrically through a lawn of grass. Avenues were usually as wide as the house and sometimes wider. They were the entrance used by carriages. Alleys were narrower lanes.

William Byrd had used avenues of trees at Westover in Virginia early in the eighteenth century (see Plate 8). Reflecting on gardens of the eighteenth century in 1806, an American garden writer noted, "Straight rows of the most beautiful trees, forming long avenues and grand walks, were in great estimation, considered as great ornaments, and no considerable estate and eminent pleasure-ground were without several of them."[16] In 1773 Philip Fithian wrote of Robert Carter's Nomini Hall in Virginia: "Due east of the Great House are two Rows of tall, flourishing, beautiful, Poplars . . . these Rows are something wider than the House & are about 300 yards Long . . . These Rows of Poplars form an extremely pleasant avenue, & at the Road, through them, the House appears most romantic."[17]

Landowners usually dictated that the avenue leading to the entrance façade of the dwelling be wider than subsidiary ones, and often they were broad enough that the entire front of the house was visible from the far end. Usually a 200-foot-long avenue would be about 14–15 feet wide, a 600-foot-long avenue about 30–36 feet wide, and a 1200-foot-long avenue about 42–48 feet wide. Now and then garden designers manipulated the perspective, so that the apparent size of an avenue

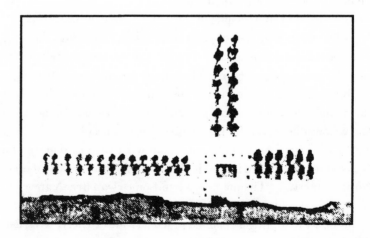

A plan of Westover, across the James from Williamsburg, 1701. William Byrd used tree-lined lanes to make his house the focal point of perpendicular axes. William Byrd II Title Book, Virginia Historical Society.

was lengthened by gradually narrowing the width of the avenue toward the far end.[18]

Occasionally, landowners planted their avenues before they began construction of their house. In 1753 a plantation owner received a letter from a relative in Scotland asking him to write and "by all means mention the fine Improvements of your garden & the fine avenues you've raised near the spot where you'r to build your new house."[19]

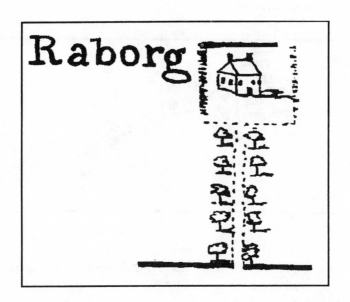

Often, avenues extended into the countryside and terminated with impressive vistas. In 1762 Hannah Callender wrote of William Peters' Belmont near Philadelphia, "A broad walk of English cherry trees leads down to the river. . . . One avenue gives a fine prospect of the city. . . . Another avenue looks to the obelisk."[20] Avenues of cherry trees were common on plantations in Pennsylvania at that time.

The grounds surrounding the country seat of Christian Raborg, a French merchant who arrived in Baltimore during the Revolution, boasted a dramatic straight avenue of evenly spaced trees leading to a simple dwelling on a property with no other formal gardens or ornamentation. At most large Chesapeake country seats of the period, however, avenues and alleys of trees were only one component of more complicated garden plans, such as those at Major Thomas Yates's house in Baltimore.

Gardeners also used smaller alleys of trees, to help define their gardens. Consisting of single or double rows of trees or hedges, these alleys usually bordered walkways, as in Rogers' grounds. Alleys through the center of a garden were wider than intersecting ones. Occasionally the designers also manipulated the perspec-

tive of these alleys, so that their apparent size was lengthened, by gradually narrowing the width toward the far end. Often, colonial gentry used the term *alley* to
refer to the walkways that ran between beds of plants and were bordered by low-
growing shrubs. Chesapeake gentlemen used their garden alleys to offer cooling
shade and exercise, direct the line of sight, define garden compartments, and add
ornament to their grounds. More often than not, they planted fruit-bearing plants
as their alleys. George Washington planted "Apricots and Peach Trees which stood
in the borders of the grass plats."[21]

In 1800, a visitor to Henry Pratt's estate, The Hills, later renamed Lemon Hill,
near Philadelphia, recorded this description: "Mr. Pratts garden for beauty and
elegance exceeds all that I ever saw—It is 20 rods long—and 18 wide An alley of
13 feet wide runs the length of the garden thro' the centre—Two others of 10 feet
wide equally distant run parallel with the main alley. These are intersected at
right angles by 4 other alleys of 8 feet wide—Another alley of 5 feet wide goes
around the whole garden, leaving a border around it of 3 feet wide next the
pales."[22]

Garden planners continued to define garden spaces by outlining individual
beds and squares with borders, often of fruit trees and bushes. The garden of
Charles Norris in Philadelphia was, "laid out in square parterres and beds, regularly intersected by graveled and grass walks and alleys . . . with . . . roses inter-
mixed with currant bushes, around its borders." Near the end of the century,
flowers were once again popular. Henry Pratt's garden was composed of twenty
squares, each with a 3-foot-wide border, "the border of every square . . . decorated with pinks and a thousand other flowers."[23]

Shopkeepers and artisans stuck with practical plants, often using vegetables in
their borders. In Williamsburg in 1786 Joseph Prentis "Sowed Lettuce Seed, on
Border on left Hand under small Paling in the large Garden . . . Sowed Lettuce
on small Border under Yard Pales." In Annapolis in March of 1792, William Faris

"Sowed a border next the Dining Room with Radish & Large Winter Cabbage." And he planted sage around his ornamental statue.[24]

Because the majority of country seats were self-sufficient units, the grounds usually contained several practical auxiliary buildings and work yard areas in addition to the main house and its geometric ornamental and utilitarian gardens. The majority had a springhouse, a milk house, a smokehouse, and a stable. Many of the grounds also contained barns, sheep houses, cow houses, pigsties, icehouses, washhouses, root cellars, poultry houses, and summer kitchens; and a few had greenhouses, stovehouses, chaise houses, and bathhouses as well.[25]

On smaller properties, homeowners often divided the land closest to the rear of the house into a woodyard or a stackyard for storing straw and a fenced family yard, which served as a barrier against potential domestic and wild animal intrusions.[26] In his *Essays and Notes on Husbandry and Rural Affairs,* John Bordley wrote that the family yard should be planted in clean, closely cut grass and that its margin alone should be allowed to contain purely decorative flowers. The well often stood near the family yard and wood yard. Sheep houses and pigsties commonly had their own individually fenced yards, and many poultry houses, or coops, had a distinct poultry yard covered with fresh sand and gravel. Sections devoted to animals usually had watering troughs within their yards.[27] Women did the washing and ironing in washhouses, which were usually within or near a separately fenced area where the wash was hung on lines or spread across shrubs to dry. Contemporaries called these areas "bleach yards."[28]

Few gardens in the Chesapeake incorporated into their design the game parks that were found in Europe. In Great Britain, the hunting of deer and other game was the legislated province of the rich and was conducted on private deer parks. The concept of deer parks expanded into landscape parks in the later part of the eighteenth century. British landowners planted their landscape parks in patterns that encouraged the nesting and increase of a variety of game. In America, deer and other game roamed free and were relatively abundant. In the Chesapeake, almost everyone was allowed to hunt deer, rabbits, and game birds. Gamekeepers and poaching were not necessary.[29]

Not only were deer plentiful in the American countryside, but they were also destructive when confined and took space away from agricultural pursuits. In 1727 Virginia's governor, William Gooch, decided that he could turn the large park at the Governor's Palace in Williamsburg "to better use I think than Deer." Daniel Fisher noted the absence of deer in the William Penn Proprietor's Garden in Philadelphia. George Washington's Mount Vernon did have a deer park, where English fallow deer and wild deer ran through the thickets. Washington received gifts of deer from friends and well wishers, as he did rare plants. George Mason's Gunston Hall, not far from Mount Vernon, also had a deer park. Mason's was stocked with domesticated native deer. The few deer parks that did exist in early America served social and ornamental rather than practical ends.[30]

North of Baltimore, at Hampton, Colonel Ridgely planned his estate to be self-contained, in order to support the day-to-day activities of his family and numerous workers. Hampton's springhouse was similar to others dotting Baltimore's hillsides. Springhouses, and the cold baths that sometimes accompanied them, needed a constant supply of cool water; the temperature of spring water at its source is normally in the mid-fifties Fahrenheit in the Chesapeake. The cool water would circulate in channels beneath the brick or stone floor level of the well-ventilated building. Often there was little dry floor area, but the largest portion of area was devoted to the spring water channels, in which crocks were placed to cool. At Hampton's springhouse, the fresh water spring emerged from under a decorative Gothic stone arch, circulated through the container channels, and left through an outlet purposefully made small, to discourage entry by unwanted animals.

A second practical structure on Hampton's grounds was the icehouse, located on the north side of the mansion. This particular icehouse was, however, more elaborate than most. Hampton's ice vault was a turf-covered mound with a domed brick ceiling, fieldstone walls, and an underground vaulted passageway. The central cylindrical chamber was almost 34 feet deep. Ice, cut from nearby ponds in the winter, was packed in straw and stored for use during the heat of summer. Chesapeake milk houses usually had a nearby supply of ice, which was used to harden butter as it was taken from the churn and worked on a slab of cold marble inside the milk house.[31]

As the nineteenth century dawned, some Chesapeake garden designers familiar with the natural grounds movement advocated the addition of artificial lakes. One of these was Rosalie Stier Calvert of Riversdale in Prince George's County, Maryland, who pointed out that her beautiful ornamental lake supplied ice for food preparation and fish for the table. On December 10, 1808, she described it to her brother: "A lake just finished which looks like a large river before the house on the southern side gives a very beautiful effect, and furnishes us at the same time with fish and ice for our ice house."[32] She had fled to Maryland from Europe with her wealthy parents in the 1790s, had lived at the terraced Strawberry Hill plantation near Annapolis as a child and later at the Paca House. Rosalie Stier married George Calvert, the son of a governor of the state. She designed the gardens and grounds at their estate with the help of her father in Belgium and a variety of professionals.

Although the privileged Rosalie Calvert was concerned with balancing beauty and utility, she carried it to an aristocratic extreme. In the carefully landscaped grounds on one side of her house were some of the plantation's slave cabins, which she designed to look like small rustic huts, complete with quaint thatched roofs. She even styled one slave cabin like a small temple with classical columns.[33]

More practical Chesapeake landowners, including Thomas Jefferson, had for decades been adding smaller, less ostentatious ponds to their grounds and stocking them with fish. Fishponds were popular throughout early America. On November 12, 1748, Peter Kalm, a European traveler, wrote, "Not only people of rank, but even others that had some possessions, commonly had fish ponds in the country near their houses. They always took care that fresh water might run into their ponds . . . for that purpose the ponds were placed below a spring on a hill." John Beale Bordley advised that no trees except willows should grow near the fishpond, "as fallen leaves and rotten wood are pernicious to the fish; as is water running from hemp, dunghills, stables, and wash houses." Those fish were there to eat.[34]

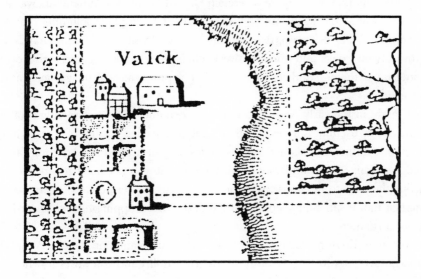

A superb example of balancing utility and beauty was the Baltimore estate of Adrian Valeck, a merchant from Holland who arrived shortly after the Revolutionary War and was named Dutch consul in 1784. Contemporary observers described his 31-acre property, called Harlem, as "a large garden in the highest state of cultivation, laid out in . . . walks and squares bordered with espaliers . . . the greatest variety of fruit trees . . . from the best nurseries in this country and Europe . . . a grove and shrubbery or bosquet planted with . . . the finest forest trees, odoriferous & other flowering shrubs" (see Plate 9). These accounts also mention a fenced kitchen garden, a greenhouse, two hotbeds with twelve moveable frames, and on an eminence a pavilion, under which was "a well-constructed ice vault." The main house, gardener's house, and stable for 7 horses and 12 cows were all of brick, the stable and carriage house of frame construction. A dairy "laid in marble" and a "pidgeon" house completed the property.[35]

An expensive brick wall surrounded the garden at Adrian Valeck's, signaling that he was a man of means and also serving a practical purpose. His gardener espaliered fruit trees along the brick wall, which absorbed the sun's heat and brought Valeck's fruit to ripeness weeks earlier than his neighbors' fruit on unprotected trees standing exposed to the whims of the elements. Colonials often called espaliered fruit trees and shrubs "wall fruit."[36]

Most gardeners could not afford brick walls and chose traditional paling—a picket fence—to protect their kitchen gardens. Occasionally Virginians denoted property lines with rail fences constructed in a zig-zag form. One traveler wrote in 1777, "the New Englanders have a saying, when a man is in his liquor, he is making Virginia fences."[37]

When the Baltimore estate of French Consul General d'Annemours was sold in 1800, it was not the elegant obelisk honoring Christopher Columbus that sold the property. The estate was described as "beautifully situated" with "a handsome grove of lofty oaks and an extensive kitchen garden and orchard well stocked with fruit of the best and choicest kind."[38] The practicality of the garden was the attraction.

Clearly, in the early American Republic, ornament was consistently counterbalanced by usefulness. In early America, if you needed a grove of trees as a setting for a statue, you planted sugar maples. If you needed a border for garden beds, you mixed currant bushes with the roses. If you needed a row of trees to define the road to the house, you chose cherry trees over poplars. There was an implicit moral sanction keeping garden design from tipping too far toward the purely ornamental.

When an English visitor wrote of one Maryland plantation, he observed, "the adjacent grounds are so judiciously disposed that utility and taste are everywhere."[39] These were the pretty and practical gardens of conservative new republicans.

II

MEANS

4

Seeds and Plants

Before the Revolution, most well-to-do Chesapeake gardeners depended on England for many of their seeds and plants, which the gentry often ordered through British agents in trade for tobacco or other goods. Shopkeepers and craftsmen could buy seed when it arrived in a general cargo shipment from England. Both groups depended on trading plants and seeds with others to keep their gardens growing. Even the wealthy William Byrd of Virginia wrote in 1721, "I went to see the Governor to beg that he spare me some bulbs for my garden."[1] Shopkeepers and gentry alike also collected alluring specimens for their gardens from the surrounding woods and meadows.

The way seeds and plants were sold changed dramatically in the Chesapeake after the war. The growth of urban economies gave rise to new commercial gardening ventures, nurseries and seed stores, operated by professional gardeners who initially imported and then grew their own seed and plant stock. Some of these professional artisans also retained staffs to lay out and maintain the gardens of others under contractual agreements. Even rural gardeners, who had traditionally traded seeds and plants with neighbors and friends on both sides of the Atlantic, began to patronize capitalistic seed merchants and nurserymen. These businessmen aggressively advertised their wares to a public increasingly concerned with the status of the ornamental, in addition to utilitarian, aspects of their gardens. Leisure time was growing in the new nation, and both the gentry and their less wealthy neighbors could now find the time to indulge in an avocation such as pleasure gardening.

From mid-century on, gardeners close to Philadelphia could take advantage of local seed and nursery businesses. Although many believe that David Landreth (who came to Philadelphia in 1781 and opened his nursery and seed business in 1784) was the first dealer in the area, he was not. One of the earliest was a gardener named James Alexander, who sold vegetable and herb seeds imported from London in Philadelphia in 1751.[2]

From the 1750s through the 1770s the most successful Philadelphia seed merchant was a woman, Hannah Dubre (sometimes spelled Duberry), of the North-

ern Liberties area two miles from the Philadelphia city limit, on the Wissahickon
Road. She and her husband, Jacob, owned 33 acres, which they later increased to
50 acres, with a bearing orchard of grafted fruit trees, some meadow land, a large
brick house and detached brick kitchen with a pump just outside the door, a barn
and several other outbuildings, and a large kitchen garden that included many
asparagus beds. Even after her husband's death in 1768, "the widow Dubre" kept
her gardens and business going. From 1754 through 1775 she offered locally grown
seeds and fruit trees on both a retail and a wholesale basis. She warranted her
seeds as "fresh and good" and sold large quantities to local shopkeepers for resale
to their clients and to exporters for trade out of the country. Before 1770, she kept
agents in town, including John and Samuel Bissell, John Lownes, and Ann Powell near the Work House on Third Street, to supply both retail and wholesale customers who did not want to travel the two miles out of town to visit her plantation. After 1770 she used James Truman, a butcher and meat curer in Elbow Lane
near the Harp and Crown Tavern, as her city agent. By 1766 she was advertising
that she could fill large orders for "Captains of Vessels" for exportation to the West
Indies "on the shortest Notice."[3] Over a twenty-year period, Hannah Dubre expanded her operation from a small local seed concern to a large-quantity supply
business catering to merchants and international traders.

The independent seed dealers and nurserymen converging on the Chesapeake
area immediately after the Revolution were often immigrants rather than natives
and European rather than English. A Frenchman, Peter Bellet, is the first commercial seedsman and nurseryman who appears in Maryland's written records
specifically offering garden plants and seeds directly to the public. Bellet's evolu-
tion from itinerant seed peddler to economically successful nursery owner typi-
fied the general trend of commercial seed and plant marketing during this period.
Beginning as a traveling seedsman based in Philadelphia, where he operated a
seed store, Bellet eventually settled in Williamsburg. He continued to advertise
throughout the Chesapeake after his relocation to the old Virginia capital and
sold plants to Maryland and Virginia gardeners for twenty years.

As a traveling seed salesman, Peter Bellet advertised in Baltimore in December of 1785 that he was visiting the French section of the town for a brief period
and had for sale an extensive variety of flowers and seeds "not known before in
this country." Bellet was lodging in the heart of the town, at The Sign of the
Lamb tavern on Charles Street, where he offered prospective customers a printed
catalogue listing the names and colors of his bulbs, which were imported from
Amsterdam. He carried more practical kitchen garden seeds with him as well.
Bellet's first Maryland advertisement reflected the preference for Dutch flowers
among the middle and upper economic groups in the early republic. Bellet also
brought with him "elegant artificial flowers and feathers suitable for the Ladies."

Bellet called himself a "florist and seedsman" on this trip and advertised his flowers as "rare and curious."[4]

Bellet's stock became more expansive during successive selling trips. On a journey through the Chesapeake almost ten years later, in early 1793, Bellet advertised roots and seeds "collected from Europe," and he offered to send orders to Europe as well. At this point, Bellet was still based in Philadelphia and had entered into partnership with another European seedsman, M. Kroonem. They were also promoting stock that was more difficult to move from place to place than seeds and bulbs, such as trees and shrubbery, and had begun cultivating their imported European seed in Philadelphia soil. Bellet was offering a surprising number of varieties of flowers, especially roses, for sale. Bellet and Kroonem called themselves "florists, seedsmen, botanists, and gardeners" and, as Bellet had done earlier, advertised their extensive plant varieties as "curious." On this selling trip Bellet traveled his usual loop from Philadelphia to Baltimore to Richmond and back. In Richmond he took lodging at Hyland's Tavern, where he again had on hand a free printed catalogue of his stock for prospective clients. To earn enough to support himself, Bellet also hired out to graft and inoculate trees and lay out flower gardens at reasonable rates. His partner, Kroonem, remained in Philadelphia to mind the store and tend the nursery garden.[5]

In the bustling new capitol, Richmond, Peter Bellet had competition for the gardening business. In the spring of 1791, Southgate's General Store advertised fresh, imported garden seeds.[6] The next spring, a seed dealer named Minton Collins was importing flower roots and seeds from London and offering them for sale at the Shot Factory, at Richard Denny's store near the market house, and at James Dove's on the main street.[7] In the fall of 1792, Collins consolidated his stock at Denny's store and had imported new seeds and flower roots to sell to his growing clientele. By the next spring, he had collected enough capital to open his own shop, devoted solely to garden stock. Collins' Seed and Flower Store sat on the north side of Main Street between the post office and the bridge over the James River. He sold retail to the general public and wholesale, or at least "upon moderate terms," to country shopkeepers from surrounding Virginia communities.[8] By the turn of the century, Collins was also receiving seed from the northern states and had customers in Richmond, Norfolk, and Portsmouth.[9] Another businessman, George French, appeared on the scene in 1798, importing seeds from London for sale in nearby Fredericksburg.[10] The competition in the Richmond and Fredericksburg area may have nudged Peter Bellet to look for a more permanent and lucrative base of operation.

Apparently, on one of his trips to Richmond Bellet ventured east to Williamsburg and found the quiet of its ordered streets and gardens a great relief from the mud and hassle of Philadelphia and Baltimore. In late 1793, he dissolved his partnership in Philadelphia and moved to a 5-acre plot in Williamsburg.[11] Bellet was not the town's first seed merchant. Gardeners Thomas Crease and James Nicholson, who worked consecutively at the College of William and Mary from 1726 until 1773, supplemented their income by selling seeds and plants grown in the college's botanical and kitchen gardens, as did James Wilson after 1779.[12] When James Stewart, a dyer and weaver, had returned in 1775 from eighteen months in England, he had offered seeds and roots of dye plants for sale to his fellow Virginians with instructions on their cultivation and the manufacture of dyes for linen, cotton, and woolen fabrics.[13]

When Peter Bellet settled in Williamsburg, he immediately expanded his stock and began referring to himself as a nurseryman, and from that point on, he ceased proposing to lay out and tend the gardens of others. In the winter of 1799 he advertised from his property on Gallows Street, now known as Capitol Landing Road, that he was still selling imported flower bulbs.[14]

Bellet quickly fit into the Williamsburg community. Local gardener Joseph Prentis was one of his early customers. Prentis's brother-in-law, Peter Bowdoin wrote from his plantation, Hungars, asking him to purchase plants for a friend at Bellet's nursery and offering to expedite the transaction: "My Boat will go up to the Capital Landing for the purpose of bringing a number of Trees from Belletts." Bowdoin also asked Prentis to give him plants from his personal garden but added, "if you have not as many to spare as will make fine beds, supply the deficiency from Belletts."[15]

Joseph Hornsby, who lived in Williamsburg in the Peyton Randolph house from 1783 until 1796, purchased from Bellet before he moved to Kentucky. When he decided to move, he began gathering up seeds and small plants from his own garden and sorting them into labeled bags. Those plants that he could not easily remove, he purchased from Bellet to start in Kentucky. In his Kentucky "Diary of Planting and Gardening," in March 1798, he reported that he had sown the seeds from Bellet, and "the Plants were very fine."[16]

Throughout the Chesapeake, gardeners were also selling bulb stock out of their gardens to meet the growing appetite for flowers among their neighbors. Henri Stier, a well-to-do neighbor of William Faris, like Faris, sold tulips and other bulb plants to fellow Annapolitans by opening his garden at full bloom in the spring, so that buyers could mark with notched sticks the varieties they wanted dug from the ground after the blossoms and leaves had faded away in the heat of summer.

Bellet used the same technique to sell his flower stock. He appealed to the immediacy of the senses rather than the memories of his prospective customers, in the days before color-illustrated advertising. Bellet also offered flowering shrubs and ornamental trees as well as more practical fruit trees and vegetable seeds. In the fall of 1799 Bellet's newspaper advertisements listed prices for the first time, and they noted that he was still importing seeds and plants from London. The notice promised that he would prepare a new catalogue for potential clients in the coming spring.[17] By 1799 Bellet had also added grafted fruit trees to his stock.

Bellet's next public notice appeared in October 1800. Now permanently settled in Williamsburg, his business was developing into a regional nursery and seed distributorship. How seeds, trees, and shrubs were shipped to Cheasapeake gardeners who placed orders was not specified in his newspaper notices. By 1800 Bellet was collecting and saving seed from his own Virginia beds and offering them for sale to the public in addition to his usual imported seed stock. He offered to sell seed by the pound or by the box.[18]

Bellet expanded his base of operations southwards to Norfolk. He completed his 1801 catalogue during the slow winter months of 1800 and offered it to prospective buyers at the store of a French merchant named Bonnard, at the Norfolk Market. Bellet advertised in the fall of 1801 that he had 8,000 growing trees for sale plus his usual supply of flower and vegetable roots and seeds.[19]

By 1803, Bellet's stock of fruit trees at this Williamsburg nursery had grown to 20,000, and he had regular sales agents in both Petersburg and Richmond who would accept orders for seed and plant stock. His agents in nearby towns were given their own supply of free printed catalogues. In an 1803 advertisement Bellet offered to sell his trees wholesale, retail, and on credit. So large was his stock that he was proposing to supply "country stores" with seeds and plants for resale "on the most moderate terms." Store owners intrigued by the idea could apply to Bellet directly at this nursery in Williamsburg or to his Richmond agent, said the ad.[20]

By 1804, Bellet had increased the size of his 5-acre nursery by buying 15 acres of adjoining land.[21] Here he planted even more trees, but apparently his health and energy were beginning to fail. After ten years in Williamsburg, Bellet decided to return north. In the winter of 1804, he offered for sale his 20-acre nursery of "well-manuered" land plus his gardening tools, eight slave gardeners, and livestock. By now his stock of fruit trees had grown to 100,000, but he had allowed his seed supply to dwindle to only "a small quantity," and he had bought no new perishable stock. Bellet's intention was to sell his stock, slaves, and tools before May 1, 1805, or put them all up for sale at public auction on that date, after which he planned to sell any remaining plant stock "on lower terms than usual" and then move to New York State. Orders for any part of the property or the whole could be left with Bellet's agents in Richmond or Petersburg. Bellet had sold 5 acres of his nursery and was attempting to dispose of his last two slave gardeners when he placed his final newspaper notice two winters later, just before he died in Williamsburg. Itinerant seed huckster Peter Bellet's astute marketing tactics had expanded his Chesapeake business from a nursery of a few seedlings to 100,000 trees in little more than a decade of residence at Williamsburg.[22]

A second professional nurseryman from Europe, a German immigrant named Philip Walter, arrived in Maryland in 1786, a little over a year after Bellet's first advertisement appeared. Walter wanted no part of the traveling life. He was a serious gardener who yearned to tend the land all year round. Walter was determined to begin his American business venture as a settled commercial nurseryman specializing in orchard plants. He decided to sell his products near the busy Market House at the foot of Belvedere, the elegant estate of then-colonel John Eager Howard. Townspeople came to shop at the market on Wednesdays and Saturdays, when neighboring farmers would load up their wagons with produce to sell and journey to Howard's Hill. With an establishment at the market, Walter figured, clientele would be drawn continually to his location, and they would be inspired to new heights in gardening by the awesome example of Colonel Howard's parklike gardens and grounds. Walter first advertised in the spring of 1787, calling himself a seedsman and a nurseryman, but he concentrated on selling primarily orchard stock. Twenty years after arriving in the bustling port town, Walter was robbed and murdered at his nursery, on Hookstown Road.[23]

While some European seed merchants and nurserymen such as Walter and Bellet decided to settle down and grow their stock in Chesapeake soil, others continued to import and travel. In the spring of 1790, John Lieutaud, a gardener and florist from France, passed through Maryland selling seeds, roots, and bulbs imported from France and Holland. Lieutaud used much the same method of operation as his fellow Frenchman Bellet did on his Chesapeake selling rounds. Lieutaud, who was from the province of Dauphiny, also offered the "curious" a printed catalogue. He boarded at the home of Captain Gould, on busy Charles Street in Baltimore, where potential customers could come to pick up a catalogue and, he hoped, buy seeds. To supplement his income Lieutaud proposed to prune, graft, and inoculate trees "at a moderate price."[24]

The next European seedsman and nursery owner to appear in Maryland records was Maximillian Heuisler, an immigrant from Munich, Bavaria. While Heuisler settled permanently in Baltimore, he often made day trips to neighboring towns, such as Annapolis, to meet local gardening enthusiasts and to hawk his wares. He was a regular seed supplier to William Faris. Heuisler personally delivered both plants and seeds to his Chesapeake customers. His wife never knew whether her husband would return from these trips with cash, new plants, or baskets of food: Heuisler traded for new seeds and plants to expand his varieties and stock, he sold for cash, and he accepted produce in trade.[25]

Heuisler's first advertisement as a commercial seed vendor appeared in 1791.[26] Aggressive in advertising his wares, he was always looking for new ways to attract potential customers. He paid to have his advertising notice in the February 1795 issue of a Baltimore newspaper illustrated with a woodcut of potted plants. His

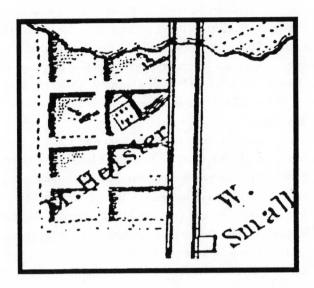

Maximillian Heuisler, seed and plant merchant, came to Baltimore from Bavaria in the 1790s. His home and nursery shared a site north of Baltimore on the Philadelphia Road.

nursery, situated on 40 acres about 1¼ miles north of Baltimore on the Philadelphia Road, was depicted on an 1801 map of the town. He regularly advertised an extensive assortment of trees and shrubberies, both useful and ornamental, for Chesapeake "plantation," orchard, kitchen, and flower gardens, plus fresh garden seeds of every description.[27] Heuisler was thought by one contemporary to be the best professional gardener in Baltimore at the end of the eighteenth century.[28] In 1803 Heuisler sold his Philadelphia Road nursery and established one closer to his Annapolis market, on the Portland-Ferry-Branch, near the southwest corner of Baltimore.[29] Maximillian Heuisler died in 1816, but his son, Joseph A. Heuisler, carried on his father's determination to build and maintain a well-respected seed and nursery business throughout much of the nineteenth century.[30]

At least two Maryland immigrants who became professional nurserymen near the turn of the century began their careers in America as gardeners under contract to busy gentlemen who had planted elaborate gardens for both food and status. Each of these gardeners saved enough capital to become successful nursery owners as the new century dawned. One was a French immigrant, John Bastian, who had come to Baltimore to supervise the elaborate gardens at Harlem, owned by Adrian Valeck. The other was James Wilkes, who had been apprenticed as a gardener in England then immigrated to oversee the gardens of George Grundy at Bolton, where he worked for three years. When Wilkes went into business for himself in 1798, he continued to offer his services as an independent gardener, available by the day, month, or year. To further supplement his income, he worked as a part-time nursery gardener for Heuisler. By 1803 he had amassed enough capital to buy Heuisler's Philadelphia Road nursery when the Heuislers opened the nursery in southwest Baltimore. Wilkes sold fruit trees and a large variety of ornamental shrubbery, greenhouse plants, and seeds imported from London, the same stock that had been the basis for Heuisler's business at that location. From 1803 until the 1820s, Wilkes sold vegetables, flowers, and exotic hothouse and greenhouse plants from his nursery.[31]

John Bastian arrived in Maryland before 1790 and was still working as gardener for the estate of Harlem in 1802. By 1808 he had begun his own independent seed and nursery business near Baltimore, and it continued until 1839. Even after his contract to tend Harlem had ended, Bastian augmented his income by tending gentlemen's grounds and gardens. Just as many of his European colleagues did, Bastian offered a full range of services to the Chesapeake gardening public, from designing to planting to "repairing."[32]

The most successful nursery business in late eighteenth-century Maryland was operated by Englishman William Booth and, after his death, his wife, Margaret. They began in business around 1793, with the sale of imported seed at two locations. Booth advertised in a Baltimore newspaper in April of 1793 that he was lodging at the home of Thorowgood Smith, Esq., in downtown Baltimore, and offering garden seeds imported from London. His second location was at Bowley's Wharf, at the harbor, where local shopkeeper acted as his agent.[33]

By May of 1794, Booth had accumulated enough capital to lease a house, and he moved next to one of the town's best-known citizens, Dr. James McHenry.[34] Although Booth did not locate near one of the town's busy farmers' markets, the house was just a half-mile west of Baltimore town, and his choice of location was a clever one. The popular McHenry had been George Washington's surgeon during the Revolution and was instrumental in developing the Constitution afterwards, so travelers and neighbors often stopped to pay their respects to him. In fact, the sociable McHenry organized regular fox hunts from his grounds into the surrounding countryside.

Initially Booth sold only seeds, which he imported from London. He worked tirelessly on the grounds and his stock during the summer, fall, and winter of 1794 and by spring of 1795 was ready for broader ventures. He placed a large notice in a local newspaper informing the public of his intention to establish a permanent nursery and seed shop on his premises adjoining the property of McHenry, with whom he had negotiated a long-term lease. McHenry's land and now Booth's new shop and home were located on the road leading to the "Federal City" and to busy Frederick town. Also, access by road to the traditional Annapolis market was easier from the south side of Baltimore than from the north or the waterbound east side.[35]

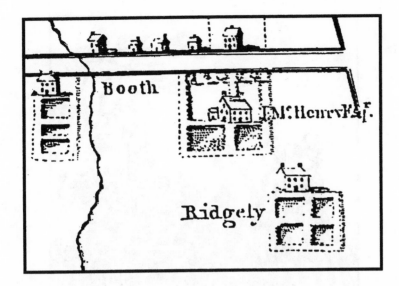

Booth had leased not only the land but also McHenry's greenhouse and had bought all of McHenry's hothouse plants, which he decided to offer for sale in pots. The surgeon had raised plants for medicinal use as well as botanical interest. Booth had a grand design for these potted plants, and he advertised it in a Baltimore newspaper. He proposed that the ladies of the town and its environs ornament their interiors with these and other potted plants during the summer months, return them to Booth for care over the winter (for a slight fee), and receive them the following spring in "full perfection." He had not only come up with an ingenious method for continuing to gain income from the plants after selling them, he planned to expand his clientele by appealing to ladies and suggesting that they use plants to decorate the interiors of their homes, traditionally the realm of women.[36]

During this same period, Grant Thorburn, a grocer who would become a famous Atlantic coast seed dealer, was drawn less intentionally into the world of ladies and plants. Until 1801 he had operated a grocery store, where he also sold flower pots. "About this time," he later wrote, "the ladies . . . were beginning to shew their taste for flowers." To make his pots more attractive, he painted some green and set them in a window. They were so popular that the following spring he added geraniums to his green pots, and from that point on he gave up the grocery business to become a seed and plant merchant.[37]

Serendipity played less of a role in William Booth's promotion of garden enterprises. Booth's capitalistic brain had been working relentlessly during the winter of 1794–95. That spring he simultaneously announced a plan to carry on a kitchen garden business that would supply specific customers with the fresh vegetables of their choice by the week, month, or year. The concept of planting pre-

chosen vegetables to supply produce on a contractual basis to his clients was an inspired version of the traditional truck farming of the region. It was Booth's clever attempt to control both supply and demand. Booth continued his original line of business, selling seeds he imported from London, as he launched his nursery, greenhouse, and kitchen garden ventures in 1795.[38]

Booth was soon the most successful professional gardener in Maryland. His training in Britain had been sound. The visiting English agriculturalist, Richard Parkinson, reported that Booth had been a gardener for the Duke of Leeds before his arrival in America. Booth's attempts to appeal to a broader market were apparently successful. In September of 1799 he advertised to the public a huge collection of "rare exotic" plants, raised in a greenhouse in cooperation with other seed and plant dealers in Philadelphia and New York.[39] The advertisement also linked the name of William Booth with some of his well-known Atlantic seaboard colleagues, David and Cuthbert Landreth in Philadelphia and David Williamson in New York, who were acting cooperatively as agents for the sale of this large collection. Also arriving in Philadelphia at the turn of the century was Bernard M'Mahon (of whom more is said in Chapter 10).

By the end of the eighteenth century, enterprising plant and seed dealers were successfully spurring on ever-widening circles of clients to new heights of interest in plant collecting and in emerging botanical class and order delineations. They also persuaded their customers that greenhouses and stovehouses were status symbols. Their sales pitch was definitely aimed at those who would see plant collecting as a reflection of their superior taste and knowledge.

In addition to his many other gardening pursuits, William Booth designed and planted some of Baltimore's most famous gardens, including the terraced falls at Hampton and those at Solomon Birckhead's Mount Royal. Booth's 1810 seed and plant catalogue is the earliest one remaining from the period in Maryland and

CATALOGUE

OF

Kitchen Garden Seeds and Plants; Physical Seeds and Plants; and Seeds to Improve Land;

FRUIT TREES & FRUITS;

ANNUAL, BIENNIAL AND PERENNIAL FLOWERS;
HERBACEOUS PLANTS AND BULBOUS ROOTS;
FOREST TREES, FLOWERING SHRUBS
AND EVERGREENS;

GREEN-HOUSE AND STOVE PLANTS.

SOLD BY

WILLIAM BOOTH

NURSERY AND SEEDS-MAN, FREDERICK-TOWN-ROAD, HALF-A-MILE FROM

BALTIMORE.

══════

CATALOGUES *to be had at the* Nursery, *where all Orders will be executed with the greatest puctuality and dispatch:* Also, *at Mr.* GEORGE ACKERMAN's *No. 9, High-street,* Old Town, PAMPHILION's HOTEL, Fell's-Point, *and at* GADSBY's INDIAN QUEEN TAVERN AND HO-TEL, *Baltimore-street,* Baltimore.

══════

G. DOBBIN AND MURPHY...PRINTERS.

....

1810.

lists hundreds of plants for the kitchen garden, sweet herbs, medicinal ("physical") plants, "seeds to improve the lands," fruit trees, annual flowers, biennial and perennial flowers, "herbacous plants," bulbous roots, forest trees, flower shrubs, evergreens, greenhouse, and "stove plants," including "a great variety of new and elegant sorts."[40]

Nineteenth-century Maryland historians claimed that William Booth was among "the earliest botanists, florists, and seedsmen in the United States" and that "his own grounds . . . were celebrated for the care and exquisite cultivation with which they were kept."[41] Booth's nursery was depicted on the 1801 Warner and Hanna Map of Baltimore. When Booth died in 1818, his inventory recorded a diverse stock, which were being made available to the Baltimore public at his seed store and at his 5-acre nursery. His widow, Margaret Booth, continued to operate the seed store and nursery through the 1820s.[42]

Chesapeake gardeners at the end of the eighteenth century did not depend solely on seed merchants and nurserymen for their seeds and plants. In fact the gentry and the middling sorts alike were still using traditional techniques of exchanging plants. Wealthy Charles Carroll of Carrollton wrote from Annapolis to friends in England for seeds he remembered from his years of British schooling. While the Carrolls continued to buy seeds from London and the colonies, the elder Carroll instructed his son as to which neighbors would give him seeds and starts from plants he admired.[43]

During the same period, Annapolis craftsman William Faris both bought and traded seeds and plants. On March 3, 1792, he noted in his diary, "Planted Carrots and parsnips that Mr. Wallace sent me for Seed"; and on May 5 of that year he wrote, "Doct Scott sent Me Some Carnation or rather pink plants & I sent him some Evening primrose plants." Faris traded for or received as gifts most of his garden plants and seeds, as did the majority of gardeners at the turn of the century. When Faris did buy seeds and plants from Baltimore, he sometimes sent cash for the garden stock by way of ship Captain John Barber, who ran a regular shuttle between Annapolis and Baltimore. Faris recorded in his fiscal accounts on March 7, 1798, "Cash sent by Capt. John Barber to Mr. C. Robinson for garden seeds—7/6." Usually, however, Faris bought his Baltimore seeds from Maximillian Heuisler, who personally delivered them to Annapolis. The capitalistic nursery and seed business was nipping at the heels of traditional garden barter exchanges.

Some gardeners still ordered their stock directly from England, especially the gentry, like the Carrolls, who had been ordering goods from Britain through their factors for decades. Faris's neighbor, Dr. Upton Scott made a list of flowers from the English garden periodical *Curtis's Botanical Magazine* and recommended to the Edward Lloyd family, at Wye plantation on Maryland's Eastern Shore, that "if cultivated at Wye, [they] would add greatly to the beauty and elegance of that delightful Place." Scott advised the English dealer, "It is hoped the Nurseryman employ'd will endeavour to execute this Commission with fidelity & dispatch . . . under an assurance that, if he transacts the Business satisfactorily, he will have more calls upon him from this quarter of the Globe."[44]

But direct orders to England diminished as early Chesapeake seed merchants and nurserymen began to offer a wide variety of seeds and plants, both imported and locally grown, to the public. They could appeal directly to potential customers' senses, by selling flowers at the height of their bloom, and to status seekers who were amassing plant collections, by offering unusual stock. They also tailored their sales promotions to the changing gardening market in the region, as it expanded beyond traditional gardeners, who planted principally for sustenance, to those who planted for pleasure and status during their growing leisure time, decorating both house and grounds with plants. Gardening for pleasure was no longer just the province of a few wealthy planters but increasingly an avocation of the expanding class of artisans and merchants, who were amassing capital that they could exchange for ornamental luxuries that would proclaim their status to their neighbors. In the early years after the Revolution, these emerging groups were continually coaxed by clever entrepreneurs to dispose of their extra capital on ornamental gardening.

5

Laborers

Surprisingly, Chesapeake landed gentry and small town merchants and arti-sans employed the same kinds of help in the garden during the latter half of the eighteenth century. William Faris used apprenticed and indentured white servants, free and slave blacks, and his own family to maintain his Annapolis garden. At the homes of the gentry, the family seldom helped with garden tasks, except that the wives usually managed the daily activities of the garden, as well as the house staff. Before the Revolution, most wealthy landowners also employed indentured whites and free and slave blacks as their primary garden staff. Later, they began to hire independent professional gardeners.

Independent professional gardeners plied their trade much earlier in public gardens. While laborers in one form of bondage or another maintained most private Chesapeake gardens, the church and state often used independent professionals to supervise their planting efforts well before the War for Independence. In Virginia early in the century, a succession of professional gardeners who were not serving under an indenture worked at institutions of the royal government in Williamsburg, including the Governor's Palace and the College of William and Mary. Some of these professional gardeners held pristine credentials. James Road, an assistant to George London, Royal Gardener to King William and Queen Mary, came to Virginia in 1694 to collect plants for shipment back to Hampton Court Palace and probably to lay out the earliest gardens at the college. He was followed by Englishman Thomas Crease, who supervised the gardens at both the college and the Governor's Palace from 1726 until he died in 1756. James Nicholson, who was born in Inverness, Scotland, sailed to the colony in 1756, to garden at the college, remaining in the position until his death in 1773, at which point he was earning the unusually high salary of £50 a year. James Wilson began as college gardener in 1773, after a politically unsuccessful tenure as palace gardener from 1769 through 1771, and he managed to serve as head gardener until 1780. The royal government appointed its first native-born Virginian, Christopher Ayscough, to the post of head gardener at the palace in 1758. When he left the post in 1768 he was earning only £20 annually for his labors. Immigrant English gardener James

Simpson briefly replaced Ayscough at the palace for £16 a year, but either the low wages or the high humidity caused him to beg to return home a scant year later. John Farquharson, a Scot, served as the palace's last head gardener, supervising the slaves, who did the daily labor, from 1771 until 1781, when the palace, by then a military hospital, was destroyed by fire.[1]

Members of the clergy were interested in both the ornamental and the practical aspects of gardening; between 1739 and 1765, Father Arnold Livers, a Jesuit priest who was raised in Maryland, kept lists of both his kitchen garden plants and the flowers grown in the parish gardens, as part of his official church records.[2] Sometimes the church employed professional gardeners. The Society of Jesus occasionally paid independent garden contractors to maintain their kitchen and medicinal botanical gardens. In May 1741, Father James Whitgrave, at Newtown, Maryland, hired William Hues as gardener, his payment to be partially in cash and partially in tobacco.

Private Chesapeake landowners, who did not have the benefit of public or church monies, usually relied on the cheaper labor of indentured servants and slaves to install and maintain their gardens. In 1765, eight years after moving to Annapolis, William Faris began in earnest to plan his grounds, just as his gentlemen neighbors were turning their attention to ornamenting theirs. On August 8, Faris contracted gardener William Jennings from England to work under an indenture, to assist him in transforming the town lots that sat beside and behind his house into gardens for both use and beauty. After four years, Faris apparently had developed his town property to such a point that it could be maintained by less skilled labor, so on March 3, 1769, he placed a notice in the *Maryland Gazette* offering to sell the remaining time of his servant gardener.

Chesapeake landowners commonly rented the unused time of their indentured gardeners to others. The practice of renting out servants and slaves with special skills allowed those who could not afford to buy the indenture or the slave to have an opportunity to use their expertise in the planning and installation of their gardens or to undertake special projects without a large capital outlay. Occasionally, landowners simply lent idle or unfriendly garden servants to family and friends. In the spring of 1751, in Williamsburg, John Blair Sr. (1687–1771) lent to Peyton Randolph (1722–75) his gardener, of whom "Mrs. Randolph gave a fine account." The servant had a history of picking fights with Blair's slaves, and in the end, apparently Blair valued his slaves more than his feisty gardener. Shortly after the servant's return, Blair "ordered the gardener to go, for I couldn't bear him."[3]

Even the wealthy rented the services of others' skilled workers when they undertook extraordinary projects. Charles Carroll of Annapolis rented two servant gardeners in 1770. He wrote to his son, "I will give Colonel Sharpes Gardener £3 per month computing 26 Working days to the Month & I will allow the Man who

Works with Him 40/ per month if He be a good Spadesman." When these particular rented servant gardeners arrived in Annapolis, Carroll was less than enthusiastic, "Mr. Sharpes . . . Gardener . . . I do not like His looks as they are very Scottish, He may buy Rum."[4]

By the time the younger Carroll returned to Annapolis after completing his
education in Europe, in 1765, the grounds had already been set in some order.
The Carrolls had begun working on their gardens in 1730, with the assistance of
a servant gardener. Later, when planning the extensive renovations of their property, the Carrolls decided to buy the indenture of a 22-year-old Welsh convict gardener, in addition to renting the two gardeners from Colonel Sharpes. Over the
next few years they directly employed several indentured servant gardeners as
well as renting other artisans. In 1771 the Carrolls used indentured servants and
slaves to dig drainage ditches. (The gentlemen themselves were busy ordering
seeds, grasses, and clover from their English factors.) In 1772 various laborers built
garden gates and a washhouse, and by 1774 brickmasons had laid the brick wall
surrounding the gardens. Stonemasons and slaves completed a sea wall at the bottom of the garden terraces in 1775, and the next year laborers were erecting the
two octagonal pavilions that would sit 400 feet apart at either end of the sea wall.
The servants' and slaves' final addition to the garden, a bathhouse, was up and
working in 1778. The artisans and gardeners who achieved these complicated additions to the Carroll grounds at Annapolis worked side by side with Carroll slaves
regularly assigned to garden work. By the 1780s, the Carroll garden was established and only needed to be maintained, so after that date the Carrolls employed
few new white garden servants, using for the maintenance work the slaves who
had been trained during the renovation.[5]

Similarly, young Annapolis attorney William Paca had traveled to England in
1761 to further his legal training. Shortly after his return, he married wealthy Ann
Mary Chew, on May 26, 1763, and began to plan his Annapolis home and gardens, which he began building in 1765. Paca employed at least one indentured
garden servant, who doubled as a shoemaker, to help plan and construct his brick-
walled pleasure grounds, which were dominated by geometric terraced gardens
that fell to a small naturalized wilderness garden area boasting a pond, a Chinese-
style bridge, and a classical pavilion (see drawing on p. 24).[6]

The vast majority of pre-Revolutionary Chesapeake gardeners of whom there
are records were indentured and convict servants from Scotland, Wales, Ireland,
and England. Although slaves often assisted in the gardens during this period, their
tasks or trades were usually not recorded, so it is difficult to verify their numbers.
Originally, most indentured servants imported into the Chesapeake worked in the
labor-intensive task of raising of tobacco. The second half of the eighteenth century, especially in Maryland, witnessed growing urbanization and manufacturing

as well as a steady diversification in agriculture away from tobacco and toward less labor-demanding crops, such as wheat. White indentured and convict servants increasingly became employed in a variety of trades.

Of the thirty gardeners identified in Maryland documents before the Revolution, all but four were white indentured servants. Many were seasoned gardeners.[7] The first Maryland servant gardener appeared in Anne Arundel County records in 1720. Chesapeake colonists looked for specific experience in their indentured servant gardeners. Charles Carroll of Annapolis asked each prospective gardener "How long he served, in what Place, in what Places and Gardens He has Worked Since He was out of his apprenticesh[ip], in What Branch He has been Chiefly employed, the Kitchen or Flower Garden or Nursery, whether He understands Grafting Innoculating & Trimming." The Revolutionary War disrupted the flow of indentured and convict servants from Britain to the colonies, and between the end of the war and the turn of the century only five additional white indentured gardeners appeared in Maryland records.[8]

During the 1770s, these indentured servants were usually paid between £6 and £32 per year plus their meat, drink, washing, and lodging.[9] Garden servants often supplemented their regular duties in the winter by doubling as shoemakers, dyers, and weavers.[10] Familiarity with the dyes produced by various plants led gardeners naturally into textile trades. This combination of crafts flourished outside the Chesapeake as well.[11] Slaves who served as summer gardeners also sometimes doubled as shoemakers during the winter months.[12]

Before the Revolution, at least nine convict gardeners arrived in Maryland. A few convict servants were sent to the Chesapeake to fill a special request, because they possessed a specific skill or trade; but most were transported to Maryland docks, and then their labor was sold, much like other imported goods of the period.

Some of Maryland's convict gardeners had practiced the gardening trade before arriving in the colonies, and one possessed an unusual knowledge of sophisticated gardening techniques. In January of 1768, Charles Carroll the Barrister, cousin to the Carrolls of Annapolis and Carrollton, wrote to his English agents, "I am in want of a Gardener that understands a Kitchen Garden . . . Grafting, Budding, Inoculating and the Management of an orchard and Fruit Trees . . . under Indenture . . . for four or five years . . . There come in Gardeners in every Branch from Scotland at Six pounds a year."[13]

The requested servant arrived at Mount Clare later that year and was apparently well respected by the Barrister and his fellow gentry, even though he was a convict. When Carroll of Carrollton bought a gardener for his father at the docks in Baltimore, he asked the Barrister's convict gardener to interview the new immigrant and then wrote his father, "I have bought a new gardiner from Captain Frost. I gave £23 currency for him; he is not above 21 years of age, appears to be healthy & stout and orderly; he says he understands a kitchen garden pretty well; Mr. Carroll's gardener examined him: he has 4 years to serve."[14]

Carroll Barrister's convict gardener may have been a good judge of men, but he did have a few negative qualities. Five years into the man's indenture, the exasperated Carroll placed an advertisement in the *Maryland Gazette* on May 6, 1773: "TEN POUNDS REWARD . . . Ran away . . . a convict servant man, named John Adam Smith . . . by trade a Gardener; has with him . . . a treatise on raising the pineapple, which he pretends is of his own writing, talks much of his trade and loves liquor."

The issue of the treatise is an interesting one. In October 1770, Mary Ambler of Jamestown, Virginia, had visited Mount Clare and noted in her diary: "at the Garden . . . he is now building a Pinery where the Gardr expects to raise about an 100 Pine Apples a Year He expects to Ripen some next Sumer."[15] It is remarkable that convict gardener Smith talked with Mary Ambler about pineapples in 1770 and had a treatise on the fruit with him in Maryland. The pineapple's popularity had grown in England during the mid-1750s, creating a demand for publications giving directions for its culture. John Giles (1726–97) published the first monograph on the plant in England in 1767. Since the Barrister's convict gardener arrived in the colonies in 1768, his claim to have written his own treatise is an intriguing possibility.[16]

Most gardeners who ran away during the pre-Revolutionary years were indentured servants, not slaves, and most records of them that survive are fugitive notices in contemporary newspapers. The advertisements placed to apprehend runaway gardeners described these servants—their clothing, mannerisms, and bad habits—in hope of speedy identification and capture. Many garden servants bore the scars of health problems such as smallpox, frostbite, cataracts, and past violence. Some convict gardeners wore double-riveted steel collars as a mark of their status especially if they had a history of "stealth of self."[17]

Occasionally masters collared white indentured servants for other offenses. In 1770 one of the servant gardeners of Charles Carroll of Annapolis got drunk and insulted several women in the Carroll family. Carroll threatened to have the man whipped, but the women begged for leniency on his behalf. Carroll wrote his son, "Squires was not whipt, He wears a Collar in terrorem to others, & as a Punishment which He justly deserves, but I think to take it off soon." Carroll felt fully justified in often whipping his favorite servant gardener, John Turnbull, for drinking too much and was surprised when the man chose to work for Carroll no more when his indenture expired in 1772.[18]

Apparently some gardeners were unwittingly impressed into naval service at European ports during the last half of the eighteenth century. At least three men who ended up as gardeners in the Chesapeake region had jumped ship when they arrived in the ports of the bay, in order to get back to terra firma. One of them was 30-year-old white sailor Pierre LaFitte, who fled a French privateer in Baltimore, hoping to return to his original trade as a gardener, further inland, at Frederick, Maryland. LaFitte quickly came to enjoy some of the benefits of life on land but disliked others. He soon ran away from his gardening chores at Frederick as well; however, he did carry with him several silver spoons and a 22-year-old French-speaking black girl wearing a green petticoat.[19] None of the white indentured servant gardeners who appear in escape notices or other pre-Revolutionary Maryland records were women, but one free white woman is known to have been working as a gardener in the Baltimore area at the turn of the century.

The average age of servant gardeners was between twenty and thirty. Charles Carroll Barrister was making his request to his English factor in January 1768, he wrote, "If the above servants are Turned thirty years of age I shall like them better as they are more Likely to be Riotous and Troublesome if young."[20]

There is no record of craftsman William Faris contracting any white indentured gardeners after 1769. By this period, the clockmaker's family was growing, and his youngsters worked as garden helpers from an early age. Even during the last fifteen years of his life, Faris occasionally called on his grown children for a little help in his garden. All of the Faris children who were living close to home between 1792 and 1804 (when Faris was recording daily in his diary) helped in the garden, usually assisting a slave or hired help. Faris's unmarried sons still living in Annapolis, who had apprenticed under their father before going out on their own

Unusual caricature of a woman gardener. In Baltimore at the turn of the nineteenth century, only one white woman appeared in the city directory as a professional gardener.

as professional clockmakers and silversmiths, continued to serve as garden labor for their aging father, who was 64 years old in 1792. One son was 27 and the other was 23 in 1792.

The craftsman's unmarried daughters all helped in the garden until they left home. Faris first mentions his youngest daughter's helping in the garden in 1794, when she was fifteen. His two oldest daughters, unmarried and heavily into the Annapolis social scene, also assisted in Faris's garden in 1799, when the eldest was 25 and her sister was 24. Notation of garden work by Faris's wife, Priscilla, appears only once. In his diary Faris noted that she was usually employed at "woman's work." She fed and sewed clothing for her family and helped Faris with his need for extra hands by raising a large family.[21]

British agriculturalist Richard Parkinson and his family rented a farm in Baltimore County for several years at the end of the century before returning to England, where he wrote of his American experiences. Parkinson also noted that his children helped with gardening and farming chores but that his wife did not. Parkinson insisted that, of the help he had to hire outside of his immediate family, white agricultural and garden labor was inferior to black labor, "for all the white men I employed there ate much and worked little. . . . the black man or slave is both clothed and fed at a less expense than a white man. . . . they bear the heat of the sun much better than any white man, and are more dexterous with the hoe."[22]

William Faris owned several slaves over the period of his residency in Annapolis, and his most constant gardening companion during the 1790s was his slave Sylva. Parkinson wrote of the advantages of buying female slaves, less expensive than males, to assist in Chesapeake agricultural pursuits. He noted that they could perform gardening tasks as well as male slaves, and "the women will be a saving . . . in the first purchase, and they will wash and milk." Faris did indeed use Sylva to cook and help with general housework as well as to work regularly as his primary gardener.[23] In Virginia, George Washington employed both male and female slaves in creating his garden. A 1799 visitor to Mount Vernon wrote, "Here many male and female negroes were at work digging and carrying away ground to make a level grass plot with a gravel walk around it."[24]

Both Faris and Parkinson also rented the male slaves of others to assist with garden labor. Parkinson's writings give a glimpse into the lives of the slaves he rented in Baltimore in the late 1790s: "Though you have them slaves all the day, they are not so in the night. All the black men I employed used to be out all night and return in the morning." No matter how the rented slaves spent their nights, they were expected to begin working vigorously the following morning. Shortly after Faris employed his neighbor Mrs. Brewer's slave Harry, the craftsman wrote in his diary, "Man Harry crawled Home this morning between 6 & 7. Went to Working Dung on the older bed by the Walnut Tree."[25]

As Edward Lloyd IV designed and planted his Annapolis gardens in the 1770s, he too supplemented his indentured garden staff with slave gardeners, as had Dr. Henry Stevenson when he installed the terraced falls and geometric flat garden at his Baltimore County seat, Parnassus, in the 1760s. Pennsylvanians also used slaves as garden help.[26] In Virginia, the Williamsburg gentry vied for the services of James, the slave of Nathaniel Burwell. He served under four head gardeners at the Governor's Palace. Governors Botetourt (Norbonne Berkeley) and Francis Fauquier paid the Burwell family £12 per year for his services, and Governors Dunmore (John Murray) and Patrick Henry each paid £14 a year for his expertise. James was a master at pruning fruit trees, transplanting native seedlings, and forcing plants in hot beds and bell glasses.[27]

Black slaves were often purchased by the professional white gardeners and plant merchants who appeared in growing numbers throughout the Chesapeake after the Revolution, starting with the first commercial nurseryman, Peter Bellet, to tend their nursery gardens, sell their stock at market, and maintain the gardens of their regular clients. Professional gardener James Wilkes offered an unusually high $100 reward for his 25-year-old runaway slave gardener, John, in Baltimore in 1801. Two other independent Baltimore gardeners, Philip Walter and John Mycroft, also used slaves to assist in their gardens.[28]

Free blacks also hired on to assist with garden chores. Between 1792 and 1804 a total of sixteen free black men helped William Faris with his Annapolis garden tasks. Most were permanent free black residents of the town, but some were passing through and hiring themselves out as garden laborers for a season. In the spring of 1792, Faris hired a black garden helper, Peter Shorter. Two days later the craftsman learned that Shorter was a runaway slave, and he immediately discharged the man. Faris usually paid £12 per annum to his free black helpers. He did not specify in his diary the amount of work expected from the workers for £1 per month.[29]

By 1790 blacks composed a third of Maryland's population. In the city of Annapolis at the time of the 1800 census, out of a total population of 2,212 persons, there were 646 slaves and 273 free blacks. Between 1790 and 1800, the population of free blacks in Maryland increased about 144 percent. Slavery grew at a much slower rate.[30]

One dramatic increase in the number of free blacks occurred as a result of the slave uprising in the French colony of Saint Dominque led by Toussaint L'Ouverture. About two thousand French-speaking refugees, including well over 500 of black or mixed racial ancestry, arrived in Maryland during the summer of 1793. Faris noted in his diary, "July 10, 1793. Yesterday & too Day thare has been between 30 and 40 Vessels went to Baltimore, the most of the full of French people . . . one Vessel had near 1200 on board."[31]

After this French settlement, free black and white French gardeners-for-hire began searching for work in Maryland. These gardeners had a significant influence on Chesapeake pleasure gardening, as they introduced tropical varieties of plants and new garden designs into the region. French-speaking gardeners became so numerous that Maryland seedsmen Sinclair and Moore published their 1825 trade catalogue in French as well as English. The contributions of the French refugee gardeners from Saint Dominique were extolled by orator John Pendleton Kennedy at the first exhibition of the Horticultural Society of Maryland: "They brought with them . . . the knowledge of plants and garden stuffs. After their arrival . . . Baltimore became distinguished for the profusion and excellence of fruits and vegetables."[32]

Throughout most of the eighteenth century, indentured white servants and free and slave blacks were the backbone of the garden labor force in the Chesapeake. While free white professional gardeners and nurserymen began to appear after the Revolution in the urban areas, such as Baltimore, it is likely that, until the Civil War, most rural Chesapeake pleasure gardens in Maryland were maintained by black gardeners, some of them free and some slaves, often ill treated. On April 5, 1777, Virginia plantation owner Landon Carter, having surveyed his newly green fields and budding trees and just about able to taste the strawberries setting flower in his garden, wrote in his diary, "My gardiner now 5 days weeding his Strawberry beds and not yet half done them. They must be well whipt." As Parkinson noted, "where the livelihood is got out of the poor soil—it is pinched and screwed out of the negro."[33]

When independent white gardeners started proliferating after the Revolution, they hired out by the day, month, or year. Parkinson reported that he hired a white man in Baltimore in 1799 to mow at "a dollar a day, with meat and a pint of whisky." He also recorded the costs of labor of various agricultural and gardening tasks at different seasons of the year around 1800 near Baltimore. "Bartering in town costs one dollar and a half per day; at harvest-work, one dollar per day and a pint of whiskey."[34] Several gardeners worked at four shillings per day in the

1770s in the Annapolis and Eastern Shore gardens of Edward Lloyd IV. One of
these gardeners, James Lilleycrap, worked as a contract gardener at the Lloyds'
Annapolis garden on a daily basis during 1778 and 1779. In February 1780, he contracted to work for a full year at £300. Apparently Lilleycrap was a trained gardener hired to undertake major garden redesign and installation. He probably
employed others to assist him and paid them out of the £300. Lilleycrap's arrangement illustrates just one of the new approaches to pleasure gardening in Maryland after the Revolution.[35]

Free gardeners had been searching for work in the Chesapeake well before the
Revolution. As early as 1749, a notice in the *Maryland Gazette* announced, "James
Cook, Gardener, from England . . . performs all Sorts of Gardener's Work . . . by
the Year." Cook had come to Annapolis to garden for Provincial Secretary
Edmund Jennings four years earlier, as an indentured servant. Cook advertised
for independent work as a gardener in 1749 and 1750, but evidently he was less
than successful at finding steady employment. On November 3, 1751, Cook rein-
dentured himself as a gardener, this time to Edmund Jennings' wife, Catherine.
In 1752 the Jenningses attempted to sell the time of the indentured gardener, noting that he was "an extraordinary good Gardener . . . understands the laying out

of new work or anything belonging to a
Garden."[36]

Virginia also saw independent gardeners searching for work before the war.
In 1766, an immigrant placed the following notice in the local paper, "Lately
arrived in this colony a young man who
professes himself a GARDENER, understanding both flower and kitchen garden
. . . grafting and budding." Three years
later, George Renney, an English gardener, advertised in the *Virginia Gazette* in
Williamsburg "to undertake by the year
to keep in order a few gardens at a reasonable price."[37] Before the Revolution,
a professional English gardener named
Joseph Thompson also lived and worked
in Williamsburg, but he sided with the
Loyalists and was denounced by his
neighbors. He left the area shortly after
the war.[38]

Ads in search of independent gardeners were not frequent in Maryland and Virginia before the Revolution (they were more common in Pennsylvania). Most gardeners ordered their garden servants from British factors or contracted for them after their ship docked in port.[39] After the Revolution, gentlemen seldom sent to England for their gardeners but began to place advertisements for professional gardeners in local newspapers. Harry Dorsey Gough advertised in 1788, "I want to employ a complete gardener at Perry Hall . . . to undertake the management of a spacious, elegant Garden and Orchard." A similar Baltimore notice in 1795 pleaded for a gardener who was specifically adept at managing strictly ornamental flower gardens. In Richmond, Adam Hunter placed a notice in the local paper search-

ing for a "Complete Gardener, with or without a family (the latter would be preferred)" for his land near Fredericksburg in Stafford County.[40]

Towns such as Philadelphia, Baltimore, Annapolis, Richmond, and Williamsburg did not hold a monopoly on pleasure gardening in the Chesapeake after the Revolution. In the 1790s and early 1800s gardeners placed notices in the *Maryland Herald and Elizabethtown Weekly Advertiser* advertising a full range of services to prospective clients in Washington County and Frederick. These gardeners offered to lay out and manage greenhouses, hothouses, kitchen gardens, flower gardens, orchards, nurseries, and pleasure grounds.[41]

After the Revolution, most professional gardeners began to sell their services aggressively, through newspaper advertisements and personal promotion. One independent gardener searching for work, Luke O'Dio, wrote to President Thomas Jefferson on June 23, 1801. As proof to Jefferson that he had gardened for notable men, O'Dio stated that he had "done 2 pices of work on the Eastern shore of Marylan one for a Wm Paca Esqr. who was once Governor of this state and one for Mr. Chew near the same place."[42]

Other gardeners and nurserymen publicized themselves and their wares more subtly, by writing books on gardening. Two gardeners who lived in Anne Arundel County at the turn of the century were David Hepburn and John Gardiner. David Hepburn had been gardener at General John Mason's estate on Analostan Island

in the Potomac River, and at Cedar Park, the seat of Governor Mercer in Anne Arundel County. Cedar Park boasted a deer park, a rare feature on Maryland estates. Hepburn and Gardiner combined their knowledge with information lifted from English gardening books to write an early American gardening book, *The American Gardener,* which was published in Washington D.C. in 1804.[43]

After the Revolution, white gardeners working in port towns, such as Philadelphia, Annapolis, and Baltimore, were as likely to be European as they were to be British. Europeans were favored by some in the postwar years. One gentleman searching for a gardener in Baltimore in the 1790s wrote, "A Dutch or Frenchman speaking English would be preferred." Of the gardeners appearing in various Maryland records before the turn of the century, more than sixty appeared to be freemen working as professional gardeners, seed merchants, and nurserymen. Most were listed in the Baltimore city directories published with some regularity after 1796. These directories did not list slaves and only occasionally listed "free persons of colour" (the first of whom appeared in 1809). They did not distinguish between free and servant whites, and they were not a complete listing of all persons working in the town. Between 1796 and 1804, thirty-three independent gardeners were listed in the Baltimore city directories.[44]

The growth of professional gardening trades spawned a need for apprentices to assist the craftsmen in their daily chores while training to become the professional gardeners of the future. Young men who had apprenticed to other trades were sometimes called into garden service during the growing season. William Faris used his clockmaking apprentice as garden help on several occasions.[45]

At least three young apprentice gardeners had appeared in Maryland records by the end of the century. These boys were employed by both professional gardeners and directly by owners of private gardens. In 1788, at age 8, William Lucas was bound out to Joseph Bignall, a professional gardener. Another 8-year-old, William Martin, was apprenticed to Jacob Eichelberger at his private residence in Baltimore County in 1786, to learn the art of gardening.[46]

A lad named Cornelius Lary was bound out by his sister at the age of 10, as a gardening apprentice to John Toon, who operated Toon's Gardens in Baltimore County. Toon's Gardens was one of several commercial pleasure gardens operating in the Baltimore area at the turn of the century. In the 1800–1801 Baltimore City directory, the 10-acre Toon's Gardens was described as being "situated about two miles down the [Patapsco] river . . . on an elevated situation," and was said to "command a view of the city and bay." "During the summer months," the directory recounted, "a great concourse of citizens make excursions by land and water to these Gardens . . . with all kind of refreshments."

Occasionally gardeners advertised their need for apprentices through help-wanted notices in local newspapers. When the enterprising William Booth was establishing his nursery, he advertised, on March 2, 1795 in the *Federal Intelligencer and Baltimore Daily Gazette*, that he was looking for one or two boys between the ages of 10 and 12 to serve as apprentice gardeners. As an inducement, Booth noted that in several years such trainees would be capable of managing the gardens of gentlemen's country seats around Baltimore which, he declared, were "going to ruin, for the want of a skillful gardener."[47]

Booth's ad reflected the post-Revolutionary trends in professional gardening in the eighteenth-century Chesapeake. White indentured and convict servants from the British Isles and black slaves had toiled side by side to design and maintain Chesapeake gardens before the war with England. After the Revolution, growing numbers of immigrant European, Caribbean, and British independent gardeners, assisted by white apprentices and free and slave blacks, took over the work. The American Revolution was the turning point in the development of independent gardening trades in the newly emerging capitalist nation. However, some aspects of Chesapeake gardening did not change until the Civil War.

6

Garden Books

Eighteenth-century Chesapeake gardeners ordered their design and planting instruction books, as well as their seeds, plants, tools, and often gardeners themselves, from mother England. Despite wide discrepancies in both soil and climate among the colonies themselves and certainly between the colonies and Britain, gardeners up and down the Atlantic depended on English gardening publications until well after the Revolution. The English garden books that dominated the American market until early in the nineteenth century, however inadequate and misleading their planting instructions, are valuable tools for reconstructing not only which plant materials were recommended but also the methods that were used in designing and laying out elements of eighteenth-century Chesapeake gardens.

The catalogues of the circulating libraries that blossomed in the bay area after the Revolution are one important aid in locating gardening books used during this period. Other documents that sometimes mention garden books are letters, inventories, newspaper advertisements, and diaries. Records of the libraries of Thomas Jefferson survive, and some private book collections from the period remain in Maryland. Among these are the library of the Ridgely family at Hampton in Baltimore County and the books of Charles Carroll the Barrister, housed at Mount Clare in Baltimore City. An examination of the books read by Chesapeake gardeners may help to explain their tenacious refusal to let go of the formal garden concepts of the ancients—like the geometric terraces found at both Hampton and Mount Clare—and accept the natural grounds revolution of their English contemporaries.[1]

Although scattered estate records dating back to 1674 exist in Maryland, these documents are nearly complete for the eighteenth century, because a law was passed in 1715 requiring all executors to make an inventory within three months of death. Unfortunately, early Maryland inventory clerks were not often very specific when recording book titles and seldom listed authors, so the interpretation of precisely what book was recorded in early property lists remains difficult. *The Art of Gardening*, which appears in several early inventories, was probably the work of the English author, Leonard Meager. His book, actually entitled *The New Art of Gardening*, was published in London in 1697.[2] The 1718 inventory of William Bladen, who was Secretary of Maryland in 1701 and Attorney General in 1707, listed John Evelyn's *The Complete Gardener*, published in London in 1693.[3] It was a translation of a French work by Jean de la Quintinie (1629–1688), who was the "Chief Director of all the gardens of the French King Louis XIV." John Evelyn (1620–1706), born in Surrey, was the author of books on architecture, engraving, history, painting, politics, and gardening. Between 1641 and 1647 Evelyn visited gardens in Holland, France, and Italy. (He called the formal terraced falls he found in Italy "hilly Gardens.") During his travels, he kept an extensive diary, which was the basis for his later writings. In 1652 he settled at Sayes Court at Deptford, where he gardened in the French style. He later designed an Italian villa garden for his brother.[4]

The extant letters of eighteenth-century Marylanders, such as those of Henry Callister and Charles Carroll the Barrister, often mention gardening books. Callister (1716–65) spent several years in a Liverpool counting house before his employers sent him to their store at Oxford on Maryland's Eastern Shore. Evidence of the frequent exchange of books among gardeners on the Eastern Shore is found scattered throughout Callister's letterbooks. After his arrival in Maryland, Callister became acquainted with a prosperous planter, William Carmichael, who lived near Chestertown.[5] Callister's letterbooks show that he borrowed Carmichael's copy of Jethro Tull's *Horse-Hoeing Husbandry: Or, an Essay on the Principles of Tillage and Vegetation*, published in 1733. The book was a classic, and Tull (1647–1741) came to be called the "father of modern husbandry."

Callister's letters mention that he owned a copy of *History of Plants Growing about Paris, With Their Use in Physick, and a mechanical account of the operation of medicines, translated into English, with many additions, and accommodated to the plants growing in Great Britain*, written by Joseph Pitton de Tournefort (1656–1708), translated by John Martyn (1699–1768), and published by C. Riverington in London in 1732. Callister offered to sell this book to a fellow Maryland gardener in 1765: "I have a small posthumous work of Tournefort. . . . it gives the description & use of plants in medicine, with their chymical analysis; it is an 2v. 12 degree worth 12/6 Currency. I shall send it if you like. I would now . . . but there are a few things in it which I would read first."[6] The book's author became professor of botany at the Botanic Jardin du Roi in Paris in 1683 and later made various expeditions in Europe and the Near East in search of plants. The book's English translator was professor of botany at Cambridge from 1732 until 1762.[7]

Callister also owned the two-volume collection *Figures of the Most Beautiful, Useful and Uncommon Plants Described in the Gardener's Dictionary*, published from 1755 to 1760 in folio. The illustrator was Philip Miller (1691-1771), one of the most important English horticultural writers of the eighteenth century. An acquaintance of Callister's, hearing that he had the collection and knowing that Maryland's governor, Horatio Sharpe, owned Miller's *Gardener's Dictionary*, mentioned to the governor that Callister might sell the illustrations. Callister wrote to the governor, offering him the watercolor plates for 15 pounds, which he declared was his actual cost: "As your excellency is possessed of the Dictionary . . . they will be curious illustrations of his subject. . . . your beneficence is seen in your laying hold of the occasion to ease me of a burthersome article, for the piece is indeed costly, and your taste seems to run rather on improvements in agriculture than mere entertainment in botany and natural history. For this I sincerely thank your Excellency."[8] After this humble appeal his altruism, the governor did buy the books.

Philip Miller was the son of a Scotsman who served as a gardener in Kent before becoming a market gardener near Deptford. Young Miller assisted his father but soon went on his own as a florist, garden planner, and nurseryman. In 1722 he was appointed curator of the Physic Garden of the London Apothecaries at Chelsea, where he served for forty-eight years.[9]

Miller's *Gardener's Dictionary* was the staple of most Chesapeake garden libraries. Virginians George Washington and Thomas Jefferson owned copies, as did many Marylanders. The complete title surveys the scope of the work: *The gardeners dictionary: containing the methods of cultivating and improving the kitchen, fruit, and flower garden. As also, the physick garden, wilderness, conservatory and vineyard . . . Interspers'd with the history of the plants, the characters of each genus, and the names of all particular species, in Latin and English; and an explanation of all the terms used in botany and gardening, etc.* It was first published in London in 1731 and revised in many editions over the coming years.

A copy of Miller's *Gardener's Dictionary* still exists in the library at Charles Carroll, Barrister's Mount Clare. The Barrister's father, who came to the colony about 1715 to practice medicine, became a planter, ship builder, land speculator, and part owner of a large iron business. Like many of the other Maryland planters, the elder Carroll ordered his books directly from England. One of the first things the Barrister did after his father's death was to pay debts his father owed a London bookseller.[10]

In 1760 as Charles Carroll the Barrister began to plan his new house in Baltimore, he ordered a copy of Miller's *Gardener's Dictionary*. By 1766 he was ordering seeds from his British factors by noting specific seed types directly from the English gardening books on his shelves. That year, he copied a long list of seeds from Hale's *Complete Body of Husbandry*, first published in London in 1755–56, and asked his English agents to send as many of them as possible to Maryland. His letters also frequently referred to the *Gardener's Dictionary*, using it to describe varieties of peach and apricot trees he wished to plant in the orchard that sat on the hillside near his terraced formal garden. He wrote, "The Nursery Man may Look into Millars *Gardeners Dictionary* where he will See the Names of Each."[11] Another book referred to by the Barrister was Thomas Hale's *Eden: or a compleat body of gardening . . . (or rather by Sir J. Hill) etc.*, published in London in 1756–57. Little is known of Hale, but Sir John Hill (1716–75) was the son of a Lincolnshire clergyman and brought up to be an apothecary. During his apprenticeship, he attended lectures on botany at the Chelsea botanic garden. In 1760 he assisted in laying out the botanic garden at Kew and was a gardener at Kensington Palace. Carroll's copy of Hale's *Complete Husbandry* still exists in the library at Mount Clare.[12]

The Barrister was also interested in the agricultural reforms sweeping England. He not only knew which specific books he wanted his English agents to buy but was able to direct them to the publishing houses in London that stocked the desired works: by letter he requested a William Anderson to order *"A new and Complete System of Practical Husbandry*, by John Mills Esquire, editor of *Duhamels Husbandry*, printed by John Johnson at the monument and his *Essays on Husbandry* including *On The Ancient and Present State of Agriculture* and *On Lucern* printed for William Frederick at Bath 1764. Sold by Hunter at Newgate Street or Johnston in Ludgate Street."[13]

The Barrister's distant cousins, Charles Carroll of Carrollton and his father Charles Carroll of Annapolis, also ordered their gardening books from England. By the mid-1760s, the family library included *New Improvement of Planting and Gardening*, by Richard Bradley (1686–1732), published in London in 1726, and Miller's *Gardener's Dictionary*. Bradley's work appeared on several Maryland inventories in the 1730s. Bradley had studied gardening in France and Holland and in 1724 been appointed the first professor of botany at Cambridge. Later, the Annapolis Carrolls added to their library *Gardener's and Planter's Calendar*, by Richard Weston (1733–1806), published in Dublin in 1782. Weston was a thread hosier in Leicester; he had traveled in France and Holland as secretary of the Leicester Agricultural Society.[14]

Another popular English publication that had entered Carrollton's library by the mid-1760s was *New Principles of Gardening . . . with Experimental Directions for Raising several kinds of Fruit Trees, Forest Trees, Evergreens and Flowering Shrubs*, by Batty Langley (1696–1751).[15] Langley is probably best remembered for his architectural pattern books, such as *Builder's and Workman's Treasury of Design*, which George Washington used when designing sections of Mount Vernon. Washington employed Langley's concepts in planning his garden as well. Langley disliked the hard-edged formalism that dominated British and European gardens before the onset of the English natural grounds movement in the early eighteenth century. Although Washington retained much formal symmetry and even tightly stylized parterres as the backbone of his Mount Vernon gardens, he did add many of

Langley's suggested "natural" features, including a serpentine entrance drive, groves of trees, walled gardens, and wilderness areas.[16]

The letters of Henry Callister and the several Maryland Carroll families show that many of the books owned by gentlemen were imported directly from London in exchange for the annual tobacco shipment or goods such as iron ore. Before the Revolution wealthy planters and merchants depended on their own private libraries, often exchanging books with one another. When literate gardeners died, their books were passed to others with deliberate care. At the death of the Virginia gentleman William Ludlow in the mid-1760s, his books were offered for sale to Charles Carroll of Carrollton, who chose two gardening books from the collection, including Batty Langley's treatise.[17]

Direct trade with London booksellers gradually decreased as tobacco became less important in the economic life of Maryland and as trade was curtailed briefly during the Revolution. As a result, bookstores and circulating libraries began to appear in the Chesapeake region. Their popularity coincided with the rise of literate merchant and artisan classes. Before the Revolution, there were a few booksellers in the mid-Atlantic. William Aikman was a pre-Revolutionary bookseller in Annapolis who imported quantities of books from London for sale directly to colonial readers. In the *Maryland Gazette* of June 23, 1774, he advertised for sale "Adam Dickson, *A Treatise on Agriculture* . . . 2 vol. Edinburgh, 1770."

Several Maryland booksellers, knowing that not all readers in the new nation could afford to buy books for their personal use, expanded their businesses to offer the less costly option of circulating library services. By 1783 Annapolis had a local circulating library, which offered a few gardening books to subscribers. These included Richard Weston's *Gardener's and Planter's Calendar* (published only a year earlier) and Thomas Mawe's *Everyman His Own Gardener*, published in London by W. Griffin in 1767. Mawe, the gardener to the Duke of Leeds, only lent his name to the book, to give an air of authenticity to the publication, which was actually written by John Abercrombie (1726–1806).[18]

The largest collection of eighteenth-century gardening books in Maryland was held by the Library Company of Baltimore.[19] These books formed the nucleus of information for Maryland gardeners for many years. Several circulating libraries preceded the Library Company of Baltimore in the growing town. In the December 1780 *Maryland Journal*, William Prichard advertised that he was opening a bookstore and establishing a circulating library of 1000 volumes in Baltimore. By 1784 William Murphy opened a second circulating library, on Market Street in the center of the city. It is about the Library Company of Baltimore, the largest Maryland subscription library of the late eighteenth century, that the most is known. It had 60 original subscribers and 1300 volumes when it was chartered in 1796. By 1809, when the first catalogue was prepared, the library had over 400 members and 7000 volumes, at a time when the city's population was almost 35,000. With nearly a hundred formal pleasure gardens dotting the Baltimore hillsides, the Library Company's resources must have been welcome. Maryland gardeners were interested in practical advice about growing table produce and in information that would help them style their gardens to reflect their taste and power.[20]

Eighteenth-century Chesapeake public and private libraries contained many books that dealt with the theory and mechanics of ornamental pleasure gardening as well as the intricacies of the more practical kitchen gardening and orchard growing. But just as importantly, in both the privately owned and the circulating libraries could be found the writings necessary for the colonists to understand the philosophy of the natural grounds movement that was occurring in England. In the 1783 Annapolis Circulating Library were the collected works of Joseph Addison, including his "Essay On Taste" and "Essay On Nature and Art in the Pleasures of the Imagination" (originally published in the *Spectator*, nos. 409 and 414, respectively). Also in the circulating library's catalogue were the complete works of Alexander Pope, including his notes on gardening (which had appeared in the *Guardian*, no. 173, in London in 1713), his "Epistle to the Earl of Burlington," published in London in 1731, and his *Essay on Criticism*, published in London in 1711. Both Annapolis and Baltimore library patrons could also borrow collections of the works of Francis Bacon, including "Of Gardens," first published in his *Essays* in London in 1625, to compare with the new views of Addison and Pope. Later in the century the works of Edmund Burke, including *A Philosophical Enquiry Into the Origin of Our Ideas of the Sublime and Beautiful*, published in 1759 were available in Chesapeake libraries.[21]

One English writer, William Marshall, decided to combine two of the most important late-century observations on the English pleasure grounds movement in one publication in 1795. *A Review of the Landscape, a Diadactic poem (by Richard Payne Knight) also of An essay on the picturesque (by Sir Uvedale Price): together with practical remarks on rural ornament*, was available at the Library Company of Baltimore. William Gilpin's *Three Essays: On Picturesque Beauty; On Sketching Landscape; and Landscape Painting* could also be borrowed there. Gilpin (1724–1804) was interested in truly natural scenery rather than the stylized natural landscapes being carved out of the English countryside by landscape "improvers." His practical ideas about the picturesque were developed into abstract theory by Uvedale Price in his two volume *Essays on the Picturesque as compared with the Sublime and the Beautiful*, published in London by J. Robson in 1794, which was also in the Baltimore library. In addition to books describing the natural grounds movement, colonial gardenres had the opportunity to see and buy prints of the latest English gardens, landscape parks, and country seats.[22]

While Chesapeake colonists were somewhat influenced by the English natural grounds movement, they were more drawn to early Roman and later Italian Renaissance garden theory and design as they had evolved in European gardens. The symmetrical, ordered designs of Andrea Palladio (1508–80) significantly influenced both home and garden design in the eighteenth-century Chesapeake.[23] Langley's popular pattern books borrowed heavily from Palladio. In the few instances where Palladio depicted landscaping, he designed the space immediately around the house to continue the proportions and designs of the interior spaces. However, by the middle of the eighteenth century, Anglo-Palladianism had evolved so that the garden was designed to be a response to the natural site and to serve as a foil to the formal architecture of the building. By contrast, Chesapeake gardeners intentionally chose to make the garden a logical extension of the house, reflecting its geometry and style, so that house and garden became a series of orderly, linked rooms.

Roman works on gardening and farming, such as Marcus Porcius Cato's *De Agricultura*, Plinus Secondus's *Naturalis Historiae Libri*, the letters of Pliny the Younger translated by Melmoth, Marcus Terentius Varro's *Rerum Rusicarum Libri Tres*, and many volumes of Lucius Moderatus Columella, among them *Of Husbandry* and a book on trees, in English translation, were recorded in several eighteenth-century Maryland libraries including the Library Company of Baltimore. In the eighteenth-century Chesapeake, most personal and public repositories containing these works in their original language also housed Littleton's *Latin Dictionary*, Ainsworth's *Latin*, or Floru's *Latin and English* to assist readers in translation. Growing interest in classical gardening and farming theories spurred Adam Dickson to write *Husbandry of the Ancients*, published in Edinburgh in 1788, a book which Thomas Jefferson acquired for his library. These writings satisfied the yen for precedent, order, and restraint that appealed to a people carving a pure, new nation out of a "howling" wilderness and toiling to establish an orderly new government.[24]

When comparing the ancients with the modern English style that had emerged in full-bloom by mid-century, Chesapeake gardeners, including Library Company subscribers, pored over Thomas Whately's *Observations on Modern Gardening*. The Library Company's copy was the third edition, published in London by Payne in 1771. Whately (d. 1772) saw the English landscape planner as a painter on a colossal scale. His work, often used as a guidebook in England by foreign visitors, reflected the general English penchant for, as well as the philosophy behind, the natural style. Thomas Jefferson and John Adams carried Whately's book with them when they visited English gardens in 1784. After his tour Adams commented, "It will be long, I hope, before ridings, parks, pleasure grounds, gardens, and ornamented farms grow so much in fashion in America."[25]

While dependent upon English gardening books for practical advice as well as theory, Chesapeake gardeners were not slaves to the British trends. Eighteenth-century American gardeners made informed, conscious decisions to extend the ordered geometrical principles of design that governed their houses to the grounds surrounding them. John Evelyn's Italian "hilly gardens" dominated Chesapeake hillsides throughout the second half of the eighteenth century.

The end of the century saw the publication of Virginian John Randolph's *Treatise on Gardening*, modeled on Miller's *Gardener's Dictionary*. Randolph (1727–84) was the last royal attorney general for the Virginia colony. His is the earliest known Chesapeake book on kitchen gardening, supposedly written in 1765 but not printed until several years after his death.

Much earlier, in South Carolina, Martha Logan, a widow who ran a finishing school and sold garden plants to make ends meet, compiled a garden calendar, which was published in *Tobler's South Carolina Almanack* in Charleston in 1752 and in other almanacs for years to follow. Robert Squibb's 1787 *Gardener's Kalendar of South Carolina and North Carolina* was also printed in Charleston, but there is no record of these two pamphlets' being read in the Chesapeake in the eighteenth century. Professional gardeners David Hepburn and John Gardiner combined their knowledge with information lifted from English kitchen gardening books to publish *The American Gardener* in 1804 in Washington D.C.

Maryland agriculturist John Beale Bordley, who had lived in Annapolis and then moved to the Eastern Shore, published his 1799 *Essays and Notes on Husbandry and Rural Affairs* in Philadelphia and authored the anonymous *Gleanings from the Most Celebrated Books on Gardening and Rural Affairs*, published in 1803 in Philadelphia. Both contained information on gardening from both England and America at the end of the century.

The early nineteenth century would see the rate of publication and the popularity of American books on gardening, like Philadelphian Bernard M'Mahon's 1806 *American Gardener's Calendar* (discussed in Chapter 10), vault quickly ahead of the traditional English treatists that British Americans had pored over for ideas and advice throughout the eighteenth century.

III
MOTIVES

7

Pleasure

When fresh breezes and bright sunlight lured colonial Chesapeake families outdoors, the most ordered space they could find in the still wild and sometimes intimidating American countryside was the garden. There they gathered—to work, to walk, to think, to talk, to play, to love, and to celebrate.

The gentry painstakingly planned their gardens and grounds to provoke genteel conversations with guests. Spirited discussions of garden plants and techniques mingled with social and political gossip. A gentleman might offer his guest a glass of spirits and as they toured the garden politely boast of his "commanding" views and vistas of the surrounding countryside. Here he could show off unusual plants he had collected and exhibit his clever knowledge of garden rarities. Here he could brag that his sweet peas were ready to eat before any of his neighbors'. He could point out a new plant he had hybridized and named. He might lead guests into a labyrinth or maze he had designed, so that he could eventually rescue them from the puzzle of his creation.[1]

One Philadelphia gardener wrote that a maze should be placed so that it could be viewed from above, such as from the upper story of a house. From there the "gay and cheerful" observer could "delight in beholding others perplexed in the pursuit." But the "grave and reflecting" person might ponder that one "seems to deviate from the true path, which nevertheless conducts him the nearest way to the end" while another "though sometimes very nigh the desired object, in a manner blindfold pass by, and with every step advance on the contrary road!"[2]

Colonial gentlemen walked admiring visitors around garden paths to impress them with their skill and knowledge and control. Virginian John Custis wrote in 1734, "I am very proud it is in my power to gratify any curious gentleman in this way." Even when the plantation owner was not at home, the professional gardener might escort chosen visitors around the master's gardens and grounds. When Virginian Mary Ambler visited Baltimore, a portion of her tour of Mount Clare was guided by the owner's servant gardener. In Philadelphia, while out riding one June morning in 1797, Jacob Hiltzheimer and a friend called at Robert Morris's. Hiltz-

heimer reported that the gardener "made us some very good lemon punch, the fruit grown in the garden, and showed us a number of pineapples growing and likewise two coffee trees in bloom."[3]

Chesapeake gentlemen wanted to make sure their gardens were seen and that the layout paid tribute to them. A visitor to David Meade's country seat, Maycox, on the James River in Prince George's County, Virginia, wrote, "These grounds . . . arranged as if nature and art had conspired together . . . do honour to the taste and skill of the proprietor, who is also the architect."[4]

Many Chesapeake gentry regularly conducted business in their gardens, and some even built offices there. In 1733, William Byrd discussed potential commercial possibilities in a garden: "The colonel and I took another turn of the garden to discourse further on the subject of iron." John Dickinson built a "study" near the fishpond and grotto in his garden just outside of Philadelphia. At the end of the century, when Richard Parkinson traveled from Baltimore to Alexandria to talk business with a Virginia gentleman, he reported that they met "in a place at a distance from the house, in the garden, which he called his office." During the Revolution, when the business of the country was war, gentlemen took their conversations to the garden to ensure privacy. On July 19, 1780, Pennsylvanian Christopher Marshall (a Quaker who had been excluded from his church for his pro-war stance) walked a visitor to his garden to converse about deserters and the deployment of French troops.[5]

Gentlemen used their offices and gardens to experiment with science. They examined, collected, and compared the plants in their gardens and greenhouses. Gardeners kept careful records of the world around them: rains, snows, first buds, ripening fruits. The garden was also their outdoor laboratory for learning and applying geometry and optics. They studied the skies from their gardens.

Birdsong fascinated colonial gardeners. They listened to and recorded the varieties of birds at-

tracted to their pleasure grounds. One young southern lady exclaimed that the "Airry Chorristers pour forth their melody . . . the mocking bird . . . inchanted me with his harmony."[6] A young man, newly arrived in the Chesapeake, was struck by "the Country full of Flowers, & the branches full of birds." Colonial gentlemen fetched garden bird's nests for admiring ladies. Virginians caged red birds and mockingbirds from their gardens and even exported mockingbirds to Eng-

land for sale.[7] Occasionally southern landowners built aviaries as part of their grounds during the eighteenth century. One early-nineteenth-century visitor wrote, "The aviary . . . is filled with many beautiful birds which fill the air with their songs—the native mocking bird, canary &c. all exerting their sweet voices in a mingled harmony, and fluttering as merrily as in their native woods."[8]

Colonials often dined under the trees in their yards, to take advantage of the noises and smells and breezes their gardens offered. Pennsylvania horticulturalist William Bartram wrote, "Our rural table was spread under the shadow of Oaks . . . fanned by the lively salubrious breezes wafted from the spicy groves . . . Our

music was the responsive love-lays of . . . the gay mock-bird; whilst the brilliant humming-bird darted through . . . suspended in air, and drank nectar from the flowers."[9]

Colonials also gathered in garden houses and temples to enjoy each other's company. Charles Brockden Brown wrote in 1798 that his father's Pennsylvania garden temple was "a circular area, twelve feet in diameter . . . edged by twelve Tuscan columns, and covered by an undulating dome . . . without seat, table, or ornament of any kind. This was the Temple of his Deity." The temple's use changed dramatically over time. Brown later wrote that, by the end of the eighteenth century, "the temple was no longer assigned to its ancient use . . . This was the place of resort in the evenings of summer. Here we sang, and talked, and read, and occasionally banqueted . . . Here the performances of our musical and poetic ancestors were heard. Here my brother's children received the rudiments of their education; here a thousand conversations . . . took place; and here the social affections were accustomed to expand, and the tear of delicious sympathy to be shed."[10]

Relaxed enjoyment of such social and sensual pleasures was part of the colonial garden, but practical considerations generally ruled. Early each morning, before breakfast, the Chesapeake landowner's wife went out into the garden to survey what had ripened, and she planned accordingly the day's meals for family, workers, and guests. After breakfast, the master, often accompanied by his male guests, would mount up for an inspection of his fields, workers, and current projects, leaving "the Ladys to their Domestick Affairs." The wife was responsible for the domestic poultry and small animals, and she directed the daily work of the gardeners and food preparers. In the afternoon, a colonial lady might don one of her finer dresses and take a more leisurely stroll around the grounds she managed, often joined by friends and family.[11]

Landed gentlemen's afternoons were often "devoted to the ladys." In 1732 William Byrd wrote of walking with the ladies of the plantation, who "conducted me thro' a Shady Lane to the Landing, and . . . made me drink some very fine Water that issued from a Marble Fountain, and ran incessantly."[12]

Early American gardens were social arenas for both public and private meetings. Confidants shared secrets in gardens. Families played together in gardens. Chesapeake gentry often chose to have family portraits painted in the gardens they had created, so that friends and neighbors could admire their peaceful relationships, genteel taste, and prosperous situations for years to come.

The garden served as a wholesome place of retreat from avarice, noise, and cares. In 1762 British American Deborah Pratt wrote this poem to a friend:

> *Come Marcia, from the noisy town,*
> *To sylvan scenes repair,*
> *Where you and I'll roam*
> *And breathe the wholesome air.*
> *With earliest birds we'll cheerful rise,*
> *And adoration pay*
> *To the great Ruler of the skies,*
> *Who guides us through the day.*
> *We, in the sultry noontide hours,*
> *Will seek the coolest shade,*
> *Or, in sweet smelling jes'mine bow'rs,*
> *We'll calmly sit and read.*
> *And when with reading we have done,*
> *We'll take our ev'ning walk,*
> *To view the glorious setting sun,*
> *Or of our Friends to talk.*[13]

Both men and women returned to the garden at sundown, figuratively and literally. Occasionally gentlemen's evening promenades in the garden occurred after some hours of eating and drinking, with predictable results. Alexander Hamilton wrote of an evening at a friend's house, where the guests were "handsomely entertained with good viands and wine. After dinner he showed us his garden and parks, and . . . got into one of his long harangues of farming and improvement of ground."[14]

Generally, women in the early Chesapeake did not wander far from the immediate vicinity of their gardens and grounds. One English visitor wrote, "The

ladies . . . I have known in Virginia, like those of Italy generally speaking, scarcely even venture out . . . to walk. . . . A high situation from whence they can have an extensive prospect is their delight and in fact the heat is too great in these latitudes to allow of such English tastes to exist."[15] Garden planners tried to take advantage of all the breezes they could. One visitor to the Woodlands, in Philadelphia, wrote to her sister of its "large trees under which are placed seats where you may rest yourself & enjoy the cool air."[16]

The colonial garden was close, comfortable, and safe. Parents regularly sent their children out to play in garden grounds. Entertaining them outdoors was easier than keeping them busy inside. Youngsters raced and rolled hoops on the paths their parents walked. Schoolmasters taught in gardens. George Washington built a small schoolhouse at the end of a walk in his pleasure garden.

Although adult strolls through gardens were usually done in the company of social equals, Princeton-educated schoolmaster Philip Fithian often joined Mrs. Robert Carter III in the Nomini Hall gardens of her Virginia home. As the two walked around the garden plats, explained Fithian, they engaged in "such conversation as the Place and Objects naturally excited. We took two whole turns through all the several Walks." The mistress politely spoke of the weather and the plants with the hired tutor, stopping to give "orders" to the two "Negroes, Gardiners by Trade, who are constantly when the Weather will any how permit, working in it." Each knew his place in the master's garden. The gentleman planned his gardens and grounds so that all would acknowledge his "delicate and Just Taste," which were the result of his great "Invention & Industry, & Expense."[17]

Artisan gardeners, having less available labor, spent more of their time tending the garden. Their wives, who spent much of the day sewing, often joined them in sunlit gardens or watched from porches. Peter Kalm wrote of "porches with seats, on which during fair weather the people spend almost the whole day, especially on those porches which are in the shade . . . In the evening the verandas are full of people of both sexes."[18] Colonial women took their chores outdoors when the weather permitted. Spinners spun, and milkmaids churned in gardens. Here they met and talked with passing neighbors and strangers.

Chesapeake gardens, like gardens everywhere in the eighteenth century, served as the setting for romance. "Courtship . . . seems to be the principal business in Virginia," noted a young English merchant visiting in the 1780s. Amorous young men carved the names of their beloveds on garden trees. In Virginia, one smitten suitor wrote, "With my Pen-knife carved Laura's much admired Name, upon a smooth beautiful Beech-Tree." Hopeful admirers offered bird's nests as presents, while "impertinent" young men chased young women through garden walks demanding kisses and girls picked flowers to "try" their lovers in loves-me/loves-me-not rituals.[19]

Shady arbors, alcoves, and summerhouses were favorite meeting spots in sunny gardens. They provided privacy and seating in addition to giving both visual and structural definition to garden areas, where they usually sat at the termination of a walkway. In early America, arbors were generally open work structures with three sides, covered with flowering vines.[20] Young couples naturally took advantage of these garden nooks. Clever suitors chose the sensuousness of the garden as a setting for seduction. Then as now, gardens were sometimes the site of weddings, even large ones. A French visitor to Virginia reported a wedding with "at least a hundred guests, many of social standing, and handsome, well-dressed ladies. Although it was November, we ate under the trees. The day was perfect."[21]

Dancing a Jig

Celebrations of all sorts took place in private pleasure grounds. After the Revolution, Chesapeake friends and neighbors would gather together in sweltering July weather in gardens to celebrate the new nation's independence. One matron reported attending a Fourth of July party of more than one hundred people, held on the banks of the Potomac under a seventy-foot-long tent decorated with garlands of laurel. The table was "very well provisioned" for the garden feast. Guests were asked to come in "American made clothes," and only wines and liquors made in Virginia were served—apple and peach brandy and whiskey. "It was a completely patriotic fete."[22]

Even before they had the Fourth of July to celebrate, residents of the Chesapeake gathered throughout the warm months for outdoor fish feasts and barbecues in gardens and on grounds bordering rivers. Here the men and boys went fishing in the morning, and all the guests enjoyed the cooked fish, along with roasted pigs, with drinks in the afternoon. Young people danced to the music of fiddles and banjos.[23]

Chesapeake colonials reveled in music and dancing. All daughters of the gentry, plus the children of some merchants and artisans, received some kind of formal musical training. Music teachers and dance masters traveled from town to town and plantation to plantation teaching young and old alike. Groups of cheerful young people would gather outside to sing in the evening. Country reels and jigs were popular in colonial gardens. One visitor reported, "Betwixt the Country dances they have what I call everlasting jigs. A couple gets up and begins to dance a jig (to some Negro tune) others comes and cuts them out, and these dances always last as long as the Fiddler can play." Although colonials frowned upon elaborate court dances, old English dances with traditional tunes remained popular in the colonies. Formal partners danced minuets and French quadrilles in gardens.[24]

Gardens provided settings for quieter, more introspective moments as well. Many Chesapeake colonials retreated to the garden to think or read. Apparently, walking through a garden alone was a popular remedy for melancholia. When one young woman in Virginia chose to take her evening tour of the garden alone, she explained that she was "thinking of Home & of her Friends & indulging her fond Grief on account of their absence!"[25] Another southern lady described a long avenue, lined with two rows of oak trees, that led to a friend's house as being "designed by nature for pious meditation."[26] After a particularly trying period, wrote one Chesapeake gentleman, "I walked through the Garden several times banishing by solitude, as much as possible, reflection on Several Days past."[27]

Others chose the peace and quiet of the garden to study and memorize. Philip Vickers Fithian recorded in his journal, "Towards evening I took my hat & a Sermon, & . . . rambled about til dusk committing my Sermon to memory."[28] Pleasure readers also sought the quiet of the garden. One distracted stroller wrote, "I was walking, with a Book in my Fist, musing & stumbling along."[29] Early American painters often depicted their subjects in a garden, reading or studying. The relaxing shade of garden greenery could spur the imagination. Even a commercial tavern garden was described as having "pleasant shady bowers, where the student, or man of leisure, sheltered from the noonday sun, and inhaling the fragrance of the surrounding aromatick plants might luxuriantly roam into the realms of fancy."[30]

In the middle of the eighteenth century, colonial gardeners occasionally designed areas of their grounds in honor of a special friend or relative, often one who had recently died. Friends and family would gather there to remember and celebrate that person. The plans of one young southern lady for such a memorial included a grove of "solemn" cedar trees but also flowering plants to infuse the area with the "freshness and gayty of spring."[31]

People were also buried in gardens, places of beauty and pleasure. Groves of trees, sometimes planted around the gravesite, were popular emblems of remembrance. Weeping willow trees were recommended for planting at burial sites, to add shade, "coolness, and ornament" and because the willow was said to be "the best and quickest corrector of impure air of any tree that grows."[32] In the nineteenth century, American landscaped cemeteries would feature groves as settings for mourners and monuments, and the weeping willow tree would become a common symbol for remembrance. From childhood through death, the garden played many roles in the lives of those in the colonial Chesapeake.

PLATES 1 & 2. Bolton, Baltimore. At the time of this painting, ca. 1805, the estate had falling terraces planted in grass and a semicircular flower garden on the lowest terrace, originally planted in vegetables. The detail shows an interesting feature, probably an arbor. Arbors and alcoves created with plantings were popular garden elements. Painting by Francis Guy. *Courtesy Maryland Historical Society, Baltimore.*

PLATE 3.   Detail of Virginia scene painted ca. 1800 shows a fenced rectangular forecourt and the avenue of trees often planted to guide the viewer's eye toward the house. Artist unknown. *Courtesy Winterthur Museum (item 64.2101c), Winterthur, Delaware.*

PLATE 4. George Washington's Mount Vernon. This 1790s painting shows the fenced deer park and the home's placement high above the Potomac River. Artist unknown. *Courtesy Mount Vernon Ladies Association.*

PLATE 5.   Perry Hall, begun in 1773, near Baltimore on the road to Philadelphia, displays the common design features of a tree-lined entrance leading to a fenced forecourt. Detail of painting by Francis Guy, 1805. *Courtesy Maryland Historical Society, Baltimore.*

PLATE 6.   The garden façade of Mount Deposit, Baltimore, built 1791–93. Detail of painting by Francis Guy, 1804–5. *Courtesy Baltimore Museum of Art.*

PLATE 7.   Entrance façade of Mount Deposit, Baltimore. Forecourts were usually fenced but seldom contained anything more decorative than a balanced planting of shrubs. Detail from painting by Francis Guy. *Courtesy Maryland Historical Society, Baltimore.*

PLATE 8.   Westover, across the James River from Williamsburg, Virginia. This 1825 painting by Lucy Harrison shows the avenue of trees planted by William Byrd a century earlier. *Courtesy Museum of Early Southern Decorative Arts, Winston-Salem, North Carolina.*

PLATE 9.   Harlem, Baltimore. The gardens planted by Dutch merchant Adrian Valeck in the 1780s served both utility and pleasure. The greenhouse seen here was added later. Painting by Nicolino V. Calyo, 1834. *Courtesy Winterthur Museum, Winterthur, Delaware.*

PLATE 10.   Pennsylvania sampler showing a tree-decorated falling garden capped by an impressive house. Terraced gardens appear in many Delaware Valley samplers worked between the 1780s and the 1820s. Reprinted from Betty Ring, *American Needlework Treasures* (New York: E. P. Dutton, 1987), "Jane Shearer: Her Work, 1806."

PLATE 11.   Parnassus, Baltimore. This miniature by Charles Willson Peale depicts the estate's falling terraces edged by rows of trees and the wide road leading up to the house. Watercolor on ivory, ca. 1769. *Courtesy Maryland Historical Society, Baltimore.*

PLATE 12. Mount Clare, Baltimore. The extensive gardens of Charles Carroll, Barrister, included a vineyard and a greenhouse and stovehouse for winter housing and growing of plants. Detail of painting by Charles Willson Peale, in a private collection.

PLATE 13.   This imposing Baltimore house and its plantings were constructed in 1782 by one merchant, Paul Charles Gabriel de Ghequiere, sold in the 1790s to another, Jeremiah Yellot, who named it Woodville, then to a third, Hugh McCurdy, who named it Grace Hill. Painting on chairback by Francis Guy. *Courtesy Baltimore Museum of Art.*

PLATES 14, 15, & 16.   William Paca
Garden, Annapolis. In the background
of this portrait of William Paca, painted
by Charles Willson Peale in 1772, is seen
the lower portion of Paca's garden, with
its Chinese-style bridge (*detail*). The
garden was restored in the 1970s (*opposite*), but the only documented part is this
section painted by Peale and written
about in his correspondence. *Painting
owned by the State of Maryland, on loan to
Maryland Historical Society, Baltimore.*

PLATE 17. Dovecote on Shirley Plantation on the James River,
Virginia. Doves were both ornamental and practical elements of
the eighteenth-century garden. *Photo by author.*

PLATE 18. Kitchen garden at John Blair House, Colonial Williamsburg, is typical of eighteenth-century kitchen gardens—organized, geometric spaces bordered by edible plants or decorative privet or boxwood. *Photo copyright Karen Stuart.*

PLATE 19. Decorative garden, David Morton House, Colonial Wiliamsburg. On town properties, where the house frequently sat directly at streetside, gardens were often placed in side yards, with the most decorative sections close to the street. *Photo copyright Karen Stuart.*

PLATES 20 & 21.   Greenhouse at Mount Vernon, Virginia. George Washington's greenhouse was reconstructed to appear as it did in the eighteenth century. It is thought to be similar to the greenhouse at Mount Clare in Baltimore, because Washington consulted the plans used by Charles Carroll, Barrister. *Photos by author.*

8

Food

Clearly many Chesapeake gardeners, whether they were rich or poor, were planting for both ornament and utility. Some of the weathiest men in the Chesapeake chose to plant vegetables and herbs in the beds on their pleasure garden falls. And yet, some recent Chesapeake agricultural historians have concluded that during the eighteenth century "the most popular leaf vegetable was cabbage, and it was said, by contemporary accounts, to be the only vegetable aside from turnips regularly consumed at the American table."[1] But evidence from period diaries, fiscal accounts, travel journals, correspondence, and newspapers indicates that this was not the case during at least the last third of the century. This widespread consumption of a variety of fruits and vegetables may have improved the health of the general populace and contributed to the enhanced stability of social conditions that prevailed in the Chesapeake at the end of the eighteenth century.[2]

Especially after the Revolution, Marylanders living near the Chesapeake Bay, where the majority of the state's population was clustered, were eating a variety of fruits and vegetables at their home tables and at the local inns. Observations of contemporary residents and visitors contribute to a picture of thriving local farmers' markets, where the majority of fresh produce not raised at home was bought. The less perishable items of imported produce were available at local general stores.

The diversity of the fruits and vegetables raised by George Washington and Thomas Jefferson is well known and could have been part of the popular wave of agricultural experimentation common among well-to-do late-eighteenth-century planters and lawyers. But the diary and fiscal accounts of Annapolis craftsman and innkeeper William Faris and the observations of visitors to the state indicate that the growth and consumption of many varieties of table produce by all levels of society was widespread during the last third of the century. Using traditional barter methods, Marylanders were exchanging diverse fruits and vegetables to grow in their own kitchen gardens, as well as buying garden and orchard stock from the rapidly emerging group of capitalistic seed merchants and nurserymen.

William Faris's accounts offer a surprisingly detailed record of what produce

The Farmer going to Market

was raised and purchased by the middling sorts in the Chesapeake Bay area between the Revolution and 1804, when Faris died. Faris began his Annapolis garden in 1761. On his town plot the innkeeper grew most of the food, other than meat, consumed by his family and guests. While his diary, spanning 1792 to 1804, reveals which edible plants he grew, his 1772–1800 fiscal accounts provide a corresponding description of the supplemental fruits and vegetables he purchased at local farmers' markets.

The English agricultural writer Richard Parkinson, while visiting the Chesapeake in the late 1790s, described the cost of typical meals that he was served at inns such as the one operated by Faris in Annapolis: "The breakfast is half a dollar, the dinner one dollar, and supper half a dollar." Chesapeake taverns served fine dinners, including fruits of many sorts, which were spread on the table after the dinner cloth was taken away. Dinners at better Chesapeake taverns usually cost about twelve shillings at the end of the eighteenth century.[3]

Parkinson declared, "Vegetables are so much used in America, that young clover is very frequently eaten for greens in the spring." He reported that the stalks of turnips and cabbages were often set out in the garden for sprouts, their tops eaten both cold and boiled. "Indeed, in the spring, they boil everything that is green, for use at the table." Sprouts sold at a quarter of a dollar per peck at the local farmers' markets in the 1790s.[4]

Although he never mentioned buying sprouts, innkeeper Faris did buy a variety of vegetables at local markets. He planted leafy green vegetables in great quantities year after year but only rarely bought greens (kale, spinach, broccoli, lettuce, and collards, then called colewart) and cabbages (including brussel sprouts).[5]

The loose-headed savoy cabbage was popular in Chesapeake homes at the end of the century, and the scotch or drumhead round cabbage was in great demand. Drumhead cabbage grew to a moderate size and was saved as "sour-crout" for the winter. In the Chesapeake, the preparation of sauerkraut was simple: cabbages were shred into a barrel, salted, and left there until needed. Cabbage was also eaten fresh as salad and boiled as a hot vegetable. Parkinson reported that in the Chesapeake, "all cabbages, intended for winter, are either placed in a cellar with earth to put the roots in, or taken up and planted thick on a bed in the garden, and a shelter-like house erected over them, covered with pine-tree boughs, or thatched with straw, which causes the outer leaves to rot, and the whole of the cabbage to have a very unpleasant smell and taste."[6]

Onions also were stored over winter. They were a relatively high-priced item to buy at market, so Faris and his neighbors grew them in quantity annually. It is not surprising that onions were expensive. In the southern states, as Parkinson reported, onions took two years to raise. The first year they grew to the size "used in England for pickling." The second year they were set out in beds, where they grew to a "tolerable" size and were equal in quality to the onions in England. In the northern states onions were raised in one year; and Long Island produced great quantities, but there is no evidence that New York onions were shipped for sale to the ports of Baltimore or Annapolis.[7]

Peas and beans were eaten fresh and also left on the vine to dry for storage throughout the winter months. William Faris recorded purchasing peas and beans only occasionally, but he grew several varieties in abundance throughout the period he kept his diary. In the Chesapeake, peas and beans of all kinds, especially kidney beans, Indian peas, and bunch beans produced well and were favored at local tables. They ran to great heights, climbing on branches or sticks.

The concept of eating fresh rather than dried peas became popular in the late seventeenth century, first in the Netherlands and then in France. A passion for fresh peas swept the French court and eventually all of Europe. "The subject of peas continues to absorb all others," wrote one courte-

Peas.

san. "The anxiety to eat them, the pleasures of having eaten them and the desire to eat them again are the three great matters which have been discussed by our princes for four days past." And so it was in the Chesapeake. Charles Carroll of Annapolis wrote to his son in October of 1773, saying, "I send You . . . a larger Dish of Green Peas than the last: I gatherd a good Dish on the 24th & a very large

Dish on the 25th, 12 dined with me & all eat of them, most were Helped to them twice, yet a good Plate full went from the Table."[8]

While the Annapolis innkeeper and his Chesapeake neighbors raised a variety of common vegetables, they were also attracted to the unusual. Faris had two large quantities of cauliflower shipped to him by packet from Baltimore. In March of 1791, he spent £1.19.6 including freight for 1 barrel (2 1/2 bushels) of cauliflower shipped from Baltimore. In September of 1792, he ordered his last costly barrel of cauliflower; he began sowing it in his own garden the next spring. Faris planted his new cauliflower crop near his favorite summer vegetables, cucumbers and squash ("simlins"), which he sowed in heavy quantities year after year while purchasing extra amounts annually as well.[9]

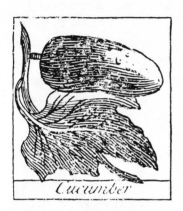
Cucumber

Faris grew cucumbers and squash in a large area of his garden devoted solely to spreading plants that needed extra room to grow and ripen. This area was also home to Faris's melon vines. Even though his garden contained various types of melons, including cantaloupes and watermelons, Faris was so tempted by the rich, moist fruit that he bought extras on occasion. Richard Parkinson wrote of Chesapeake melons, "They grow luxuriously in Virginia, and excel all the fruit I ever tasted."[10]

A related fruit, the tomato, according to recent works by Chesapeake historians, was grown as an exotic flower or curiosity rather than an edible fruit during the eighteenth century. These writers assert that the custom of eating tomatoes did not become common until the second half of the nineteenth century.[11] But a letter written in 1794 by Marylander David Bryan, who farmed for the prominent Revolutionary general Otho H. Williams, informed the general that he had bought a bushel of tomato seeds for £1 to plant as a field crop on Williams' estate in Western Maryland. Contemporary books written by Chesapeake authors reveal that the tomato was used in medicines, sauces, soups, and pickling.[12]

Celery was another popular vegetable in the Chesapeake. During the last half of the eighteenth century, Charles Carroll the Barrister and Edward Lloyd IV ordered celery seed from England to plant in their kitchen gardens. In the 1790s attorney-turned-farmer John Beale Bordley declared celery his favorite vegetable for use in soups.[13]

William Faris grew no celery in his Annapolis garden, but he regularly purchased large quantities of other vegetables often used in soups: corn, potatoes, sweet potatoes, rice, and turnips. He grew turnips himself, but no rice or potatoes,

and only a few ears of corn came from his garden.
Faris could easily purchase corn, turnips, and potatoes from farms operating near Annapolis and so
chose not to waste his limited in-town garden space
on these space-eating crops. Corn and potatoes
usually sold for between 3 and 4 shillings per bushel
during the 1790s.

Faris referred to his corn as "Roasting Ears," but
it is likely that he served corn in a variety of ways
on the tables at his inn. On September 7, 1772, one
English visitor to Maryland wrote about the corn-
based diet that was still prevalent among the poor
and on the frontier settlements. The process of preparing it remained unchanged from the days of the
colony's founding: "Indian corn, beaten in mortar
and afterwards baked or boiled, forms a dish . . .

Turnips

called hominy, and when salt beef, pork, or bacon is added, no complaints are
made respecting their fare."[14]

Chesapeake corn generally produced between twelve and fifteen bushels per
acre during the 1790s and grew to a height of 12–14 feet. White corn grew taller
than yellow, but yellow ripened earlier by a month and was sweeter. Marylanders
ate their "roasting ears" with drawn butter. They also ate corn in mush, bread,
and cakes. Mush made of corn flour was eaten with milk. Parkinson wrote, "The
better sort of people make a very nice cake, with eggs and milk, about the size
and thickness of pyfleets, or what are called crumpets in London. The lower class
of people mix the flour with water, make a sort of paste, and lay it before the fire,
on a board or shingle, to bake. . . . This is called a Johnny cake." For the corn left
to harden on the cob for winter use, "They have generally a husking-feast; when
all the neighbors come and help to husk; and, after they have done, they have a
supper, smoke segars, and drink whiskey."[15]

Most travelers were impressed by the native American corn. An account by
Richard Parkinson reads: "I went by way of Annapolis where I saw some very
beautiful Indian corn, intended for roasting ears; and, under it, cymlings, cucumbers, melons, &c. which made a most beautiful and luxurious appearance. . . .
Indian corn is the great dependence of every part of America, for both man and
beast."[16]

Root crops, such as the potatoes and turnips, which innkeeper Faris also bought
in great quantities, were a staple in eighteenth-century Chesapeake households.
One reason for their popularity was the ease with which they could be stored for
winter use. Beets, carrots, Jerusalem artichokes, parsnips, radishes, turnips, and

Aruchoke

potatoes were frequently grown in town gardens as well as on larger farms. Potatoes were the most common of these and had been in use throughout the region during much of the eighteenth century. The cluster potato was the most popular variety. The "wise" potato, or pinkeye, was light red with an excellent taste, but only a small amount of these were produced annually in Maryland. "London lords" were large potatoes generally believed to be the best potato for highyield production in the Chesapeake. Irish potatoes were preferred in the spring, when they were imported from Cork and sold at the docks in Baltimore. Irish potatoes were also grown in Maryland. Parkinson reported that there were also local varieties that took their names "from some small cause," such as the "poor-house potatoes" in Baltimore. These were summer or early potatoes, valued for taste but small in size. The harvest of Chesapeake potato crops was generally one hundred bushels to the acre in the 1790s.[17]

Another popular root crop was turnips. In Maryland, turnips were generally grown for table use, although some farmers were experimenting with feeding turnips to cattle. Some claimed production of one thousand bushels per acre in the 1790s. At the height of the harvest, turnips cost 2 shillings per bushel. In the spring, when they were a rarer commodity, they often sold for $1 per bushel, or even a penny each, although they were no larger than a hen's egg and were used in soup. Aging winter turnips sold for half a dollar per bushel. The type of turnips in general use were the old English red-top and green-top, with large tap-roots plus a few Norfolk lily-white and Norfolk green-top. Chesapeake turnips were sweet, and greatly favored among locals. Charles Carroll of Annapolis even wrote to England in 1770 for a sophisticated turnip slicer that had been recommended by English agriculturalist Arthur Young. Two years later, Annapolitan Joshua Johnson was combing London for the elusive "machine to slice turnips." Turnips were preserved for the winter in the same manner as potatoes, either put in pies or in cellars.[18]

Chesapeake householders occasionally substituted rice for potato and turnip dishes in the eighteenth century. John Beale Bordley reminisced in 1797, that he had grown rice sixty years earlier in the dry sandy soil of Annapolis as well as in the loamy soil of the nearby South River. Bordley claimed that the Annapolis rice, which was sold by the quart, was preferred to the best imported rice. He also grew rice on the Eastern Shore: "In 1781, in a clay loam on upland, in Talbot, Maryland, I grew a garden bed of it, drilled and hoed; the produce whereof was good in quality and quantity." William Faris did not, as far as we know, grow rice, but there is evidence from the 1790s that he bought it, in 3-pound quantities, paying 1 shilling for this amount.[19]

Faris did each spring plant unusually large quantities of seed for another popular root crop, radishes. Richard Parkinson confirmed the plant's popularity in the Chesapeake at the end of the century: "Radishes are much in use, and very fine, growing to the size of an English carrot." Of American carrots that Englishman did not have the same high opinion: "Carrots are raised; but they are, in general, almost tasteless, and nothing like those in England; the soil being too poor for them." Apparently most Americans did not agree

Carrots

with the visiting agriculturalist, for Charles Carroll of Annapolis took pride in the carrots he grew in that town. On May 5, 1786, the *Pennsylvania Mercury* reported that almost every family in that state grew carrots for use at their table.[20]

Cherries

Parkinson did, however, admire the Chesapeake asparagus, of which he wrote, "This country is called the West-river. . . . The first asparagus I ever saw grow by nature, was on the banks of that river, where it grows very fine."[21] Townspeople such as Faris also cultivated orderly rows of asparagus, which, combined with collections from nearby wild asparagus fields, apparently produced enough of the vegetable to satisfy their needs. The innkeeper did not record buying extra asparagus at local markets.

In fact, Faris grew many vegetables that he did not supplement with market purchases. These included beets, broccoli, brussels sprouts, carrots, colewart, kale, garlic, leeks, lettuce, parsnips, peppers, pumpkins, radishes, shallots, and spinach. He also raised eggplant for its blossoms and beautiful fruit but did not record eating the vegetable or buying it.[22]

Fruits and nuts were somewhat more costly to buy at market than vegetables, but they were also plentiful in the Chesapeake countryside. Colonial secretary William Eddis, writing home to England on September 7, 1772, related that throughout the whole province of Maryland fruit was not only bountiful but excellent in taste. There were very few farms or plantations without apple and peach orchards by the middle of the eighteenth century. Even town gardeners like Faris grew apple, peach, plum, pear, and cherry trees on their properties. Eddis reported that Maryland peach trees produced fruit of an exquisite flavor. Peach trees grew spon-

Peach

taneously in the Chesapeake, and the character of their fruit varied. Some were yellow "like a lemon," some white, and others dark-red throughout. Peaches bore

Artinc Plumb

from the stone in three years, but the fruit was thought to taste better when the tree was grafted. Peaches were often prepared as juice, in both fermented and unfermented states. Orchard fruits were so plentiful throughout the Chesapeake at this time, that Faris recorded paying for pears, plums, and apples only occasionally and did not record buying any extra cherries or peaches at market. Not all orchard fruits were equally successful in the Chesapeake; nectarines and apricots often dwindled away before they ripened.[23]

Apples did succeed in the humid Chesapeake climate and sold particularly well at the winter markets. One agriculturalist reported selling two bushels and a half (a barrel) of apples after Christmas for $5 or £1.17.6. To retain one barrel of good apples for sale at Christmas, however, required at least ten barrels, which would have been culled over time down to one. To save apples for winter, the fruit was placed in a chamber, "letting them lie there for some time to sweat." The apples were then transferred to a barrel and placed in a cellar, where they would stay cold but not freeze.[24]

Making cider was a more profitable venture, because less perfect fruit could be put to use.

Pear

.An Apple

During the 1790s cider was selling at $2 for a 3 gallon barrel. Charles Carroll of Annapolis and his son annually put away vast quantities of cider for their family and servants. In 1775, the elder Carroll put away 190 casks of "cyder" (he estimated 22,800 gallons) for the coming season. Innkeeper Faris also made gallons of cider, however he did record buying one dozen in 1800 for two pence.[25]

The Chesapeake countryside produced some very good wild grapes. William Eddis wrote, "In the woods, I have often met with vines twining round trees of different denominations, and have gathered from them bunches of grapes of a tolerable size and not unpleasant to the palate."[26] But the hope of Chesapeake gardeners in the eighteenth century was that American vineyards could one day rival those of France. Charles Carroll the Barrister, who lived near Baltimore, imported tokay grapes to plant as an experimental vineyard at his country seat Mount Clare. The tokay grapes flourished there, and their fruit was reported by contemporaries to be excellent. In the 1770s Charles Carroll of Annapolis was growing grapes experimentally in a new 4-acre vineyard planted on turfed terraces with intersecting walks. He brought from Europe two vignerons, and he planted imported claret, tokay, renish, and burgundy grapes as well as native ones from Virginia. He not only wanted to produce wine for export, but he designed his vineyard also to "be a great Ornament to the Plantation."

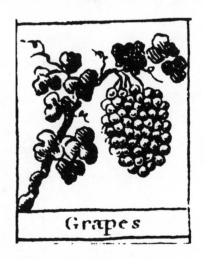

Grapes

Although his wine never reached the quality necessary for export, in 1775 he put up well over 400 gallons of "Country Wine." Even town gardeners with limited garden space cultivated grapes, which were often "tyed up & secured against the frost by straw." On March 7, 1796, Faris reported in his diary, "I've fixed up 3 posts . . . for grape vines." Grapes were eaten fresh and as juice and dried for raisins. Whether the innkeeper crushed his grapes for wine or dried them for raisins is not recorded in his diary.[27]

According to Faris's fiscal accounts between 1790 and 1800, he did buy raisins, currants, and dried plums during the November and December holiday seasons.

Faris mentioned pruning his own currant bushes on March 2, 1795. Ripe currants were bubbled into jam as well as dried.[28]

Local grocers stocked imported fruits, among them currants, raisins, dried prunes, citrons, pineapples, lemons, limes, and oranges during this period. Faris

often concocted a special punch for his guests, and his records show that he bought several large quantities of lemons and limes. Lemons usually were imported from Lisbon, while limes and oranges traveled to Chesapeake ports from Antigua during the last decade of the eighteenth century. A few wealthy Marylanders grew their own citrus fruits, but the vast majority of these exotics were bought in small quantities at local outlets or in larger amounts directly from the shippers at the docks in Baltimore and Annapolis. Unfortunately, William Faris did not record where he purchased his citrus fruit.[29]

The wife of Charles Carroll the Barrister grew her own lemons, limes, and oranges in her greenhouse at Mount Clare and generously offered to share her treasured citrus trees with George Washington after his retirement from public life. Mount Clare was also remarkable for its early production of pineapples.[30]

By the end of the century, the appeal of sweet pineapples had spread to all levels of Chesapeake society. On July 7, 1802, William Faris received "by the Packet a present from Mr. Pitt of a parcel of Pine Apples." Captain William Pitt, Faris's son-in-law, from Baltimore, was a shipping merchant who traded in the West

Indies and probably had brought a load of pineapples back on a return run. Faris did not try to grow pineapples, which were plentiful imports into the Chesapeake in the last decades of the century. One visitor noted, "Great abundance of pine apples were sold in Baltimore market, which come from the West Indies, and are retailed at a quarter of a dollar each, but are not so good as the pineapples raised in gentlemen's hot-houses in England."[31]

More delicate fruit, such as berries, were usually sold fresh at local farmers' markets. Berries added sweetness to the eighteenth-century table, where they were served in puddings, pies, breads, and with cream. Most householders also concocted some of their favorite beverages from the colorful fruits. Faris "bottled" his gooseberries as soon as he picked them from his garden vines and also grew mulberries and strawberries. The innkeeper occasionally bought extra strawberries at market, and in May of 1799 he paid 9 pence to indulge in "strawberries and milk." Faris also purchased huckleberries at his local farmers' market in July of 1800.[32]

The only nut that the Annapolis innkeeper mentioned purchasing between 1790 and 1800 was the chestnut, which he bought by the quart. Faris had planted his own almond, horse chestnut, and walnut trees; and he recorded in his diary on September 6, 1794, "Beat the Walnuts off the Tree and got I suppose a Bushel of Nuts." Parkinson wrote that there were walnut trees in great numbers in the Chesapeake but that he liked the taste of the hickory nut more. "The chest-nut tree is very handsome, and its fruit pleasant. There are but few hazel-trees; therefore, not many nuts: their shells are much thicker and kernels smaller than those in England."[33]

William Faris and his neighbors grew most of their own herbs as well, and Faris did not purchase any at market. The innkeeper's garden herbs and spices included bergamot balm, catnip, garlic, ginger, mint, nutmeg, parsley, rosemary, saffron, sage, and thyme. During the eighteenth century, herbs were the medicine of choice for most physical complaints, in addition to being used as food seasonings and air fresheners.[34]

Faris and even his wealthier neighbors used the majority of space on their town lots and gardens to produce a great variety of fruits, herbs, and vegetables for family and guests. As they did with flowers, they traded the plants and seeds of edible crops. Chesapeake travelers and residents recorded an unmistakable picture of diverse produce growing, trading, and consuming by people at all levels of society during the last third of the eighteenth century.

Society

The second half of the eighteenth century saw unprecedented economic growth in towns from Philadelphia to Norfolk. Baltimore was growing into Maryland's commercial center. Farther south, Richmond and Norfolk were expanding to meet new consumer needs, as politics and trade grew there. Newly wealthy merchants, attorneys, and shopkeepers were busily defining themselves in their stylish summer homes—country seats—with private gardens.

After the Revolution, new economic and social institutions unfolded as the fledgling nation squirmed to get comfortable with its economic and political independence. Growing urban economies supported a work force paid by the hour for their labors. Suddenly it was not just the gentry with leisure time on their hands. During this period, commercial public gardens blossomed in and near Philadelphia, Baltimore, Richmond, and Norfolk.

These commercial public pleasure gardens served as more than recreational diversions. They quickly became regular meeting places in which to conduct social, business, and political interchanges. As befit the new capitalistic fervor of American cities, these public pleasure gardens were intended to be profit-making ventures. Owners garnered money from admission fees, the sale of food and drink, and from occasional gala promotions.

Most commercial pleasure gardens offered simple amusements. Jalland's Garden, a tea garden sitting near Fells Point in Baltimore, specialized in polite social conversation accompanied by formal tea drinking and soothing music. In Philadelphia, Harrowgate Gardens, two miles north of town on the road to New York, and Grey's Gardens were the two commercial gardens for tea drinking.[1]

Men, women and children flocked to Grey's Gardens, on the banks of the Schuylkill River, during the last quarter of the century.[2] Grey's specialized in coffee and teas accompanied by "relishes." One visitor wrote of the "coffee, cheese, sweet cakes, hung beef, sugar, picled salmon, butter, crackers, ham, cream, and bread. The ladies all declared it was a most charming relish!"[3] Grey's Gardens appealed to those interested in the latest fashion in dress, food, and garden design.

At Bowling Green, near Richmond, gentlemen could enjoy an extensive array

of summerhouses and bowers for "the accommodation of company," scattered throughout the gardens. Guests could play shuffle board, ten pins, and bowls.[4] As early as 1737, the owner of Centre House tavern, on the eastern edge of Philadelphia's central square, invited "gentlemen who would divert themselves at bowls" to avail themselves of the garden and green on the grounds of the tavern.[5]

North of Baltimore, Easton's Gardens boasted a more formal garden setting, illuminated in the evenings. The most elegant eighteenth-century Baltimore pleasure garden was Chatsworth, later called Gray's Garden, which attracted men and women for innocent flirtations as well as gentlemen more interested in discussions of affairs of state and serious commerce.

By contrast, the more raucous Spring Gardens sat on the waters of Baltimore's harbor and offered intoxicating beverages and "accommodations" for gentlemen only. This garden specialized in sport—especially fishing—and attracted both the honest holiday-maker and the clever gamester accustomed to nursing pipe and glass with cards flying and dice rattling. It was a place where gentlemen, senses dulled by spirits, could lose their cares and their money with easy negligence. At Norfolk, the Wig-Wam Gardens promised bloody cock fights every Saturday night during the summer season.[6]

Also sitting on Baltimore's harbor, Toon's Pleasure Gardens offered tea and liquor as well as fishing and rural sports to both ladies and gentlemen. Amusements at large rural gardens such as Toon's would include, in addition to fishing, daytime games of skittles or nine pin, trap ball, battledore and shuttlecock, and kite flying. See-sawing and blindman's bluff were often played by both sexes, leading to predictably exciting encounters. The less physically and emotionally adventurous would gather around to watch.

Some garden operators offered bathhouses to the public at their gardens. In 1765 at New Bath gardens in northern Philadelphia, the owner advertised to "Accommodate Ladies and Gentlemen with Breakfasting, on the best Tea, Coffee, and Chocolate, with plenty of GOOD CREAM . . . and by furnishing them with Brushes and proper Towels" for their bathing pleasure. John Coyle's Wigwam Garden on the Schuylkill River and Harrowgate Spring garden in Philadelphia offered garden visitors and bathers "the best of liquors of all kinds, Breakfasts, dinners, teas, coffee and fruits of all kinds."[7] The proprietor of the Falling Garden in Richmond erected a bathing house for adults and children at his public garden. One dollar bought three baths, and two children could bathe with only one ticket.[8]

Some clever entrepeneurs sited their gardens near urban theaters, to catch the food and drink trade before and after performances. In Baltimore, one of the first public gardens sat just next door to the theater and boasted "convenient Summer-Houses" where hot tea, coffee, and chocolate, as well as "cold Colla-

tions" were served to guests by two young servant boys.[9] In Norfolk, Riffaud's Gardens sat near the local theater. On evenings when no actors graced the stage, proprietor Riffaud planned musical concerts for his guests.[10]

Some commercial gardens relied on horse racing to lure guests. In the Chesapeake, races were regularly scheduled during court days, a few days each month when events were held at the county courthouse. The Hay-Market Gardens in Richmond installed an organ to entertain patrons on race days. Garden concerts, balls, and theatrical performances kept race patrons and court visitors occupied at the Hay-Market Gardens in the evenings.[11] In Philadelphia, Hunting Park race track doubled as a public pleasure garden.[12]

Not all of these commercial garden ventures initially were designed specifically as public pleasure grounds. Spring Gardens and Toon's appear to have been tavern gardens attached as afterthoughts to favorite old sporting sites near waterside fishing haunts. Gray's Chatsworth Garden had been one of the earliest private formal gardens in Baltimore before it was converted to commercial use after the Revolution. Ironically, in this later role of the garden, the elegant classical design, with its terraced falls and eight intricate formal garden beds, lost importance.

The more formal public pleasure gardens came to life largely at night, under the illumination of torchlights or festoons of hanging lanterns strung along the sides of garden paths. The reflected swaying light lent the fragrant walks and private arbors the eerie intimacy of moving garden shapes in silhouette. The details of the living garden were lost in the dark, only the skeleton formed by the intersecting walkways separating the garden beds and the shadows remained to form a backdrop. Strolling patrons, lulled by their favorite libations, listened to the orchestra and gazed up at fiery torches glistening through the trees. Twinkling candles dotted the tables spread around the edges of the grounds. A formal plea-

Skittles

sure garden like Chatsworth served as a foil for the imagination and a stage for social and business affairs.

The same practical reason inspired the development of public gardens that had prompted the creation of the private country seats that dotted the hillsides surrounding Chesapeake towns. During the last decade of the century, many Philadelphians and Baltimoreans with money to spare built summer residences, a mile or two from the outskirts of the town proper, that they occupied from May to September. Their primary motivation was fear of the disease that plagued the teaming, muddy inner-city during the long, hot summers. One English visitor to Baltimore during the period wrote, "A tradesman is to consider another very unpleasant circumstance . . . yellow fever in consequence of which, it has, lately, been usual during three months of the year for all trades to be stopped; and he is compelled not only to lose the profits of his business for that space of time, but also to incur a certain expence . . . by providing himself with a country-house . . . or by taking lodgings for himself and family at twenty dollars per month . . . besides the danger to his own life, and the lives of all his family—a great expence for a small tradesman."[13]

During the summer months, those who could not afford to build their own country seats, temporarily escaped the sweltering city by journeying to the commercial pleasure gardens on the outskirts of the town. These public pleasure grounds were based on a concept developed by the Greeks. Athenians, too gregarious to retreat to walled Egyptian-type private gardens, planted public squares with grass and trees as sites for outdoor assemblies and teaching academies. Cicero's *Dialogues* noted that the early Roman public gardens were copies of these Greek academy parks. The concept of public gardens and parks was not new to Maryland. During the fall assembly session in 1696, Maryland governor Francis Nicholson "requested to have a Certain parcell of land in the publick pasture according to the Demencons thereof mentioned and layd down in the Platt of the Town for planting or makeing a Garden, Vineard, or Somerhouse or other use" for the new capital of Annapolis.[14] Similar areas were set aside in colonial Williamsburg and Philadelphia. By the last decades of the eighteenth century, the

concept of outdoor areas devoted to public gatherings and entertainment had
evolved into commercial enterprises for the recreation of the crowded citizenry
and for the profit of their owners and operators.

By the early 1790s, Chesapeake landowners were promoting property for sale
or rent by presenting it as opportune for development into commercial pleasure
grounds. In a Baltimore newspaper during the spring of 1793, Thomas Kelso
advertised 24 acres near the town as the ideal setting for conversion to a summer
retreat for those possessed of a taste for rural pleasures. Kelso noted that his land
was "well calculated for a public pleasure garden and the rapid increase and growing riches of Baltimore offer at this time great encouragement for such a place of
recreation."[15] Near Richmond, the French Garden, consisting of 19 acres, was
advertised for sale as a public garden. It contained 4 acres enclosed and "highly
improved" and its sitution was reportedly both "romantic" and "convenient."[16]
When Bowling Green, near Richmond, was offered for sale, it was touted as a
"desirable stand for a public house and pleasure garden, in which it is at present
occupied with considerable profit."[17] During the summers of the 1790s in towns
up and down the Chesapeake, pleasure gardens were the place to go to see and
be seen.

Maryland's pleasure gardens were modeled after those that had been in vogue
in London since the last half of the seventeenth century. Baltimore's gardens, like
those in English, were popular public meeting places and were important to the
exchange of information between gentlemen, visitors, and tradesmen. Even the
simple amusements offered at these gardens were seen as more than frivolous pas-

times. In the 1760s Charles Carroll of Annapolis wrote to his son, then studying in Europe, of the significance of such institutions in the life of a gentleman: "play houses, coffee houses, Renelaghs, Vuxhalls, Routs, Opera . . . must have allotted now & then some hours to these Genteel & necessary Amusements."[18]

Like the London pleasure gardens Ranelagh and Vauxhall, Chesapeake public gardens offered shaded bowers covered with climbing vines, to attract couples interested in serious courting.[19] Less amorous and older garden visitors relaxed in summerhouses and arbor alcoves sipping and smoking while discussing business and politics or just building a house of cards. As the elder Carroll knew, even building a house of cards could be instructive to a youth such as his son. An eighteenth-century English versifier wrote,

> *Whilst innocently Youth their hours beguile*
> *And joy to raise with Cards the wondrous pile,*
> *A Breath a Start, makes the whole fabrick vain,*
> *And All lies flat, to be began again:*
> *Ambition thus erects in riper Years,*
> *Wild Schemes of Pow'r, & Wealth, & endless cares;*
> *Some change takes place, the labour'd plan retards,*
> *All drops—Illusion All—an House of Cards.*[20]

One Baltimore pleasure garden even borrowed its name, Spring Gardens, from a famous London public garden at Vauxhall, which Johnathan Tyers, who became

the proprietor there in 1728, improved by adding an orchestra.[21] Many Chesapeake gardens offered musical entertainment for their patrons. A traditional eti-
quette was followed at formal pleasure gardens. Women came in full evening dress,
and men walked bareheaded, with their hats under their arms. A stately prome-
nade of the garden's main walks was usually first on the evening's agenda.

Patrons came from near and far. Harbors supplied visiting ensigns sporting
jaunty cockades, and Chesapeake plantation owners encouraged their trusted employees to make occasional visits to bustling cities like Baltimore and Philadelphia.
Virginia schoolmaster Philip Fithian wrote in his journal that on a visit to Philadelphia on May 20, 1774 he "walked to a lovely Garden near the Hospital call'd
Lebanon, drank some Mead, and had a most agreeable Ramble" before returning home around ten in the evening.[22]

By the end of the eighteenth century there were five commercially owned public pleasure gardens thriving in or near Baltimore. The most elegant was Gray's
Chatsworth Garden, which sat at the corner of Green and Saratoga Streets. This
old, walled private garden had been updated with the addition of serpentine path-
ways meandering around the tree-lined perimeters of the grounds. The heart of
the garden, however, remained an elegant eight-bed falling terrace garden laid
out in geometric symmetry. Mr. Mang, the proprietor in 1801 after the departure
of Richard Gray, promoted the gardens "to be extremely neat, such as forming

appearance, which may give a pleasing relaxation to the leisure hours of the industrious citizen." A band playing music entertained there during the summer months
three times a week, and accommodations and refreshments were provided for a
fee.[23]

In addition to its regular entertainments Chatsworth's Garden occasionally
held extraordinary outdoor celebrations to commemorate special occasions. A
Baltimore newspaper advertisement in September of 1794 announced an evening
of "splendid illumination" to commemorate the founding of the French republic
in 1792. Although Richard Gray invited both ladies and gentlemen to his gardens
to celebrate France's "late and glorious successes over their combined enemies,"
Gray also announced a public dinner only for interested men, to be served at the
garden on the evening of the "festive occasion." Ladies were not invited to the
supper celebration, where gentlemen of the town gathered to eat, smoke, and
toast under a blaze of light. Subscriptions for such social events often were taken
at the post office.[24]

Baltimore's pleasure grounds also served as formal political meeting places,
where the town's future was debated. In August 1794, a Baltimore newspaper announcement called on, "The inhabitants of the Precincts . . . to meet at Gray's
Garden . . . in order to take their sense respecting the incorporation of Baltimore-
town . . . on the plan proposed by the . . . last session of the General Assembly."[25]
In addition to serious civic town meetings, political celebrations were frequent in
most Chesapeake public pleasure gardens, where special entertainments were held
each year on the Fourth of July to commemorate the nation's recent independence.

During its regular entertainments and special celebrations, Gray's Garden was
usually the scene of "politeness, delicacy, and uniform conviviality"; however, occasionally rogues and "unprincipled fellows" disrupted the civility. In 1794, shortly
after Chatsworth's annual July 4th illumination and musical celebration, a notice
in a local paper reported that "a number of Lamps were destroyed and carried
off from the Garden . . . which rendered the illumination . . . incomplete." The
proprietor declared that he was outraged by this "shameful conduct" and offered
a generous reward to anyone who would "inform him of the depredators."[26]

A second, only slightly less formal Baltimore public garden was Jalland's Garden, which was renamed Staple's Gardens in 1802. These grounds were situated
between Baltimore Town and Fell's Point, a little to the east of a small stone bridge
at the foot of Philpot's Hill, which dropped to the south side of the harbor. John
Jalland's gardens were small and said to be "handsomely laid off; their central situation, salubriety of air . . . commonly serves to exhilerate the leisurely moments
of the weary visitor—while music lends its aid to soften every care. They also are
often illuminated, and always provided with the niceties of the season." A news-

paper announcement in spring of 1793 stating that Jalland had completed a "considerable addition to and improvement in his gardens" indicates that it had been opened before then. The garden was to open for the 1793 season on the evening of Friday, March 24, the announcement read, and evenings would be "regularly appropriated to tea parties, and every possible exertion" would be made to render the same pleasant and agreeable summer's recreation as in years past.[27]

Jalland's Garden was the town's most conservative public garden. Jalland offered only tea and coffee to drink, but he did provide music during evenings of "elegant" illumination and assured his guests that "no ungenteel people" would be admitted. Inclement weather caused Jalland to reschedule his annual Fourth of July celebration in 1794, but the proprietor promised his disappointed patrons that the rain-delayed garden illumination would "take place with splendor, in commemoration of a day which every tyrant must abhor, but which every friend of liberty must venerate." Jalland also vowed to provide music "suitable to the occasion" for his upcoming celebration of the anniversary of his nation's Declaration of Independence.[28]

Another formal garden in Baltimore offered more-inebriating liquid refreshments during the evenings of illumination and musical entertainment. Easton's Garden sat north of the town on the road travelers took between Baltimore and Philadelphia. Renamed Rural Felicity in 1803, the establishment was operated by Joseph Jeffers. This garden was open, torches ablaze, every night of the summer season; but on Wednesday and Saturday evenings, additional illumination was added. Jeffers prided himself on the "careful attention" he paid to accommodations for his guests, who came from all walks of life.[29]

Young people often spent their time in commercial pleasure gardens playing simple games, such as blindman's bluff; but these apparently innocent, carefree

pastimes often were seen as cover for amorous intrigues. One anonymous Englishman wrote,

> *Intent on Mirth alone the Rural Train*
> *Pass the gay vernal hours in rest from Pain:*
> *The buxom Youth hoodwink'd each other find,*
> *And innocently laugh to cheat the Blind.*
> *Thoughtless in Sport they urge the wanton Play,*
> *Nor heed the latent Pow'r that reigns in May;*
> *Beware ye tender Maids, your glowing Hearts,*
> *For Love tho' blind is not without its Darts.*[30]

Such moral maxims attended even the more boisterous garden games.

Baltimore's last two public pleasure gardens, which were sports oriented, offered
neither elegant evening illuminations nor soothing music. Spring Gardens sat
southwest of the city, down the Patapsco River, about a mile across the water from
Baltimore Town. During the 1790s the proprietor built a house on the grounds to
accommodate fishing parties. This "garden" was a place of resort and pleasant
retreat for gentlemen who were fond of angling and eager to escape the city and
women, the latter not being invited to Spring Gardens.[31] Regular summertime
amusements at sports-oriented pleasure gardens, in addition to fishing, included
leapfrog, skittles or ninepins, and trap ball. Eighteenth-century gentlemen saw an
inherent learning experince even in leapfrog:

> *While blooming Health bestows its warm supply*
> *The active Youth their Limbs elestic try*
> *By turns they yield the pliant Back prepare*
> *By turn they spring and seem to move the Air*
> *Hence learn in Life with Similar address*
> *Prudent to bend or resolute to press*
> *Your force examine ere you chuse your part*
> *The World is Leap Frog play'd with greater Art.*[32]

Chelsea, or Toon's, Pleasure Garden, which was built around 1790, was also
situated about two miles down river, but on the southeast side of town. It possessed a "delightful and captivating" view of the beautiful country seat of Captain John O'Donnell, called Canton, which had "a very handsome garden, in
great order," as well as a hothouse and a greenhouse.[33] Located on an elevated
prospect, Toon's Garden also boasted a good view of the Chesapeake Bay and

was noted for the "salubrity of its air and elegant situation." Contemporaries noted, "during the summer months a great concourse of citizens make excursions by land and water to these Gardens, where every accommodation is provided, with all kinds of refreshments." John Toon advertised in a local newspaper in the spring of 1795 that his guests could watch the "captivating" progress of the building of "the Federal frigate" in David Stodder's nearby shipyard.

During this period, Toon was attempting to improve the overland access by horse, stage, and foot to his commercial gardens, which were originally reached primarily by boat.[34] At Toon's, both ladies and gentlemen were encouraged to try their hand at fishing while enjoying the best of liquors, tea, coffee, and syllabub. But lady anglers did not dress down for the sport; quite to the contrary, they dressed in their finest to spend an afternoon fishing and hoping to be noticed. One eighteenth-century Englishman observed,

> *Silks of all colors must their aid impart,*
> *And ev'ry fur promote the fisher's art.*
> *So the gay lady, with expensive care,*
> *Borrows the pride of land, of sea, and air;*

Furs, pearls, and plumes, the glittering thing displays
Dazels our eyes, and easy hearts betrays.[35]

Public pleasure gardens in the Chesapeake ran the gamut from classic formal walled gardens to stylish fishing holes, where even the ladies could drink and be seen at sport. These commercial enterprises set the stage for the development of the free publicly planned and supported gardens and parks that urban citizens would develop in the nineteenth century, and they also were antecedents to the commercial amusement and theme parks of twentieth-century America.

Inspiration and Expression

In 1784 George Washington wrote to the marquise de Lafayette encouraging
her to accompany her husband on a return visit to the new American republic: "You will see the plain manner in which we live; and meet the rustic civility,
and you shall taste the simplicity of rural life."[1] Even though he had never been
across the Atlantic, Washington was well aware that the pleasure gardens of the
gentry in the new republic were much less sophisticated than those in England
and Europe, sometimes intentionally. But the face of the new nation was changing at the end of the century.

Immediately after the Revolution, clever European gardening entrepreneurs
immigrated to America to entice the new nationals to buy their books, seeds, and
services. They set about to create a market not only among the already pleasure
gardening gentry but among the rising merchant and working classes as well.
And they succeeded. At the end of the century, pleasure gardening was growing
among rich and middling groups alike. Ladies were becoming more interested in
decorative flowers and potted plants. The motives for this rapidly expanding interest in pleasure gardening in America were as varied as the gardens themselves.

The most important among the immigrant gardening entrepreneurs was Bernard M'Mahon (1775–1816), who came to Philadelphia from Ireland in 1796 to
establish a seed and nursery business. "He enjoyed the friendship of Thomas Jefferson. . . . the Lewis and Clark expedition was planned at his house. . . . [he was]
instrumental in distributing the seeds which those explorers collected." Contemporaries reported that M'Mahon's store was a meeting place for serious botanists
and hobby horticulturalists, a haven for artists and scientists.[2]

In 1806 M'Mahon gave America its first great gardening book, *The American
Gardener's Calendar*, which was printed in eleven editions between 1806 and 1857,
when it was superseded by Andrew Jackson Downing's *Theory and Practice of Landscape Gardening*, first published in 1841. A Philadelphia newspaper called *The American Gardener's Calendar* "a precious treasure" that "ought to occupy a place in every
house in this country." M'Mahon's main motive in writing was to expand his
profitable nursery enterprise, which supplied seeds and plants to many Chesa-

peake gardeners. Almost all of America's earliest indigenous gardening books served as the liaison between the nurseryman and an emerging middle-income group of home gardeners. As increasing leisure time and interest in the craft grew, there were not enough trained professional gardeners to go around nor funds to employ them.[3]

By 1806 M'Mahon understood the proud new country well enough to appeal to guilt and national hubris in his effort to sell his readers on the concept of pleasure gardening and thereby increase his profits. In his introduction, M'Mahon lamented that America had "not yet made that rapid progress in Gardening, . . . which might naturally be expected from an intelligent, happy and independent people, possessed so universally of landed property, unoppressed by taxation or tithes, and blest with consequent comfort and affluence." M'Mahon concluded that one reason for this neglect was the lack of a proper reference book on American gardening, a situation which he volunteered to rectify.[4]

Gardening books, plants and other supplies, and the practice of gardening itself fit into the new nation's burgeoning capitalistic fervor at the end of the eighteenth century. In addition to professional gardeners and nurserymen like M'Mahon, whose numbers grew quickly after the Revolution, nonprofessional gardeners of every stripe often sold nature's products to gain extra income. George Washington encouraged his gardener to sell extra nursery stock for a profit, one-fifth of which he allowed the gardener to keep. Nobleman Henri Stier, who had fled Belgium during the French Revolution, had a bulb sale when he moved back there from Annapolis in 1803; then he bought bulbs in Europe and shipped them to his old Chesapeake neighbors.[5] William Faris, in his fiscal account book for October 23, 1799, noted receiving the substantial sum of $40 for tulip bulbs from John Quynn. Fellow Annapolitans Alexander Contee Hanson and Thomas Harwood, and Captain John O'Donnell from Baltimore visited his garden to mark tulips and hyacinths that interested them; after the blooms faded Faris dug up the marked roots and sold, or traded, them to the gentlemen.[6] Then whole gardens became commercial enterprises, as dozens of public pleasure gardens flourished on the Atlantic coast.

But how these new Americans designed their gardens was not dictated by any capitalistic motive. Their untamed surroundings and their interest in ancient precedents prodded them toward the orderly gardens that they planted around their dwellings.[7] Early in his work, M'Mahon described "ancient gardens," by which he meant the gardens common a hundred years earlier in Britain and Ireland.[8] Ironically, just as the English were rebeling against their "ancient" geometric garden designs, M'Mahon found America's new citizens clinging to the formality of the classical past. Perhaps the young nationals were looking for the security of precedents to reinforce their present unsteady situation. The ordered and hierarchical implications of classical terraced gardens probably appealed to the gentry, who were losing their privilege of rank through association with the British and groping to maintain that privilege through natural, and therefore inevitable, order instead of through historical precedent.

Americans found the enlightened ideas of the Italian Renaissance particularly attractive during the early national era. Thomas Jefferson once boasted, "Ours are the only farmers who can read Homer." A southern lady, Eliza Lucas Pinckney, recounted in 1742, "I have got no further than the first volume of Virgil . . . to find myself instructed in agriculture as well as entertained by his charming pen." More than seventy years later, Jefferson's brother-in-law Henry Skipwith wrote to a friend who was planning an orchard, "Virgil's Georgics would have given you a full idea of his Quincunx."[9]

Library records indicate that America's literate "farmers" were also reading Richard Bradley's *A Survey of the Ancient Husbandry and Gardening*, published in 1725, and Adam Dickson's *The Husbandry of the Ancients*, published in 1788, and Columella, Virgil, Cato, and Pliny, whom they saw as providing models of day-to-day estate management, including food production.

Continental European formality dominated most Chesapeake gardens. One British visitor to the Chesapeake observed, "Frenchmen . . . appear to me to be the best judges of gardening in America, perhaps because their own climate and soil are more nearly similar to those of America, than either the English or Scotch." Even Thomas Jefferson, touted by many as a promoter of the English natural grounds movement, claimed to admire French gardening above all others. Benjamin Henry Latrobe, upon his initial tour of America between 1795 and 1798, condescendingly noted the classical influence prevalent in the Chesapeake. He wrote that the gardens at Mount Vernon were "laid out in squares, and boxed with great precision . . . for the first time again since I left Germany, I saw here a parterre, chipped and trimmed with infinite care into the form of a richly flourished Fleur de Lis: The expiring groans I hope of our Grandfather's pedantry." Americans were clinging to European gardening traditions rather than adopting the natural grounds movement.[10]

Bernard M'Mahon consistently referred to gardening as an art, just as his friend Thomas Jefferson did throughout his lifetime. As most literate eighteenth-century Americans were well aware, the educated man of the Italian Renaissance hoped to be at least knowledgeable in all of the fine arts, from painting, sculpture, and music to architecture and gardening. M'Mahon was conversant in classical letters, including literature. The Renaissance man also used the sciences—mathematics, astronomy, engineering, and botany—to shape his daily life.

Renaissance men assembled great botanical gardens, collections of plants intended for study rather than pleasure, motivated by intense interest in botany and horticulture. Many of these gardens, inspired by religious devotion, were attempts to recreate the Garden of Eden. Renaissance botanical gardens grew from the concept that if man could understand and order plants, he could understand and order his personal world and the universe at large. Early Americans searching similar paths for order and understanding, also assembled botanical collections. From early in the eighteenth century, British American colonials traded plants back and forth across the Atlantic. They grew specimen plants in their gardens and prided themselves on the rarities they could nurture. It was to these instincts that M'Mahon, the plant salesman, hoped to appeal.[11]

Under Louis XIV, the French carried to its culmination the Italian Renaissance rationale for ordering the external environment for both use and ornament. In France the concept of unifying the structure with its setting evolved into a theatrical presentation of the geometric house, balanced with a descending progression of architectural elements, such as smaller buildings, fences, gates, and steps. The great house and its dependencies were set at the pinnacle of an array of landscape features that led up to it. These designs were the work of powerful people engaged in the ultimate battle, trying to control nature. In France, complicated, controlled inert parterres outlined by clipped hedges, statues, topiary, and planned groves of trees connected the whole with the natural countryside surrounding it. Here was the ultimate unity of architecture, the decorative arts, the garden, and the natural site. The garden was meant to define and expand the image of its owner. It was this concept, intentionally stripped of most of its ostentatious excesses, that Chesapeake gentlemen adopted to help define their place in the emerging republic. In the new nation, the Chesapeake gentry used the evolving science of optics to direct the viewers' attention and to lengthen or shorten perspectives, hoping to enhance the onlooker's view of the property and opinion of its owner.

The belief that they were consciously ridding themselves of ostentatious excess was a point of honor among many in the eighteenth-century Chesapeake. M'Mahon understood this as he promoted gardens for both use and ornament. If one garden could achieve both goals, all the better. From 1745 through 1756 the weekly game of the gentlemen in the Tuesday Club in Annapolis was to mock ostentation while trying to set the colony around them into some civilized order. In an effort to explain this philosophy, Charles Carroll of Carrollton wrote to an English friend in 1772, "An attempt with us at grandeur or at magnificence is sure to be followed with something mean or ridiculous. Even in England where the affluence of individuals will support a thousand follies, what evils arise from the vanity & profuse excesses of the rich!" Only weeks later Carroll's father warned him of much the same trap: "Elegant and costly furniture may gratify our Pride and Vanity, they may excite the Praise and admiration of Spectators, more commonly their Envy, But it Certainly must give a Rationale." Both of them felt it best to "avoid any appearance of . . . ostentation."[12] Thomas Jefferson and John Adams, after their tour of English gardens in 1786, expressed similar feelings. Adams wrote, "It will be long, I hope, before riding parks, pleasure grounds, gardens, and ornamented farms grow so much in fashion in America."[13]

In American gardens a balance of useful plants and trees planted in a pleasing order was the ultimate goal. When Annapolis attorney John Beale Bordley retired to Wye Island in the Chesapeake Bay, he was determined to become a self-sufficient patriot farmer. He substituted homemade beer for spirits imported from London, kept sheep, and grew his own hemp, flax, and cotton for clothing. He knew that American dependency on Britain was drawing to a close and wrote a friend in London, "We expect to fall off more and more from your goods. . . . we are using our old clothes and preparing new of our own manufacture; they will be coarse, but if we add just resentment to necessity, may not a sheepskin make a luxurious jubilee coat?"[14] This attitude helped shrink the distance between America's landed gentry and its town merchants and craftsmen.

Another of M'Mahon's goals was to spread gardening to urban shopkeepers and artisans. Craftsmen such as William Faris, who lacked sufficient space to develop classical falling terraces but wished to copy the designs used by the grander families, adopted traditional geometric garden styles that more closely resembled the miniature formal garden adaptations of the Dutch. Land in Holland was often only a thin layer of soil unable to support the deep roots of trees, so the Dutch planted low hedges instead. As the Dutch population grew, their gardens became more compact as well. And, while the sophisticated French thought flowers too common and certainly not as controllable as the gravel they used in their parterres, the Dutch were avid flower growers. American devotion to such compact urban flower gardens could be particularly profitable for M'Mahon, who devoted much of his book to the cultivation of flowers.

Urban Americans were indeed appreciative of their gardens. In Philadelphia, Elizabeth Drinker wrote in her diary on April 10, 1796, "Our Yard and Garden looks most beautiful, the Trees in full Bloom, the red and white blossoms inter-mixt'd with the green leaves, which are just putting out flowers of several sorts bloom in our little Garden—what a favour it is, to have room enough in the City . . . many worthy persons are pent up in small houses with little or no lotts, which is very trying in hott weather."[15]

M'Mahon had come from Great Britain when English landscape architects were abandoning Western traditions of formal garden design and embracing more natural forms, which were applied to the larger expanses they called pleasure grounds. The extreme formality of French and the fussiness of the miniature Dutch flower gardens helped spark this British movement against artificiality in the garden. Influenced by writers such as Joseph Addison and Alexander Pope, whom Charles Carroll of Carrollton claimed was his favorite poet, and by the romantic landscape paintings of the French artists Claude Lorrain and Nicolas Poussin, English landscape architects rejected centuries of traditional Western garden design. English garden design reformers, such as Lancelot "Capability" Brown, and their followers favored peaceful landscapes featuring created and controlled green lawns for grazing deer and livestock, stands of needed trees, and serpentine rivers that would invite fowl and animals.[16]

The English landscape itself was to become the garden. Intricately planned serpentine rivers and lakes reflected "natural" hills planted with carefully chosen trees and shrubs. M'Mahon's treatise reviewed the use of the popular English ha-ha, a wall or ditch sunk below the level of garden, which was intended to make the gentleman's lawn appear to flow into the surrounding countryside.[17] In England, where laws kept all but the rich from hunting deer and small game, these ha-has kept the hunting preserve of the gentry secure. Ha-has were used occasionally in the eighteenth-century Chesapeake, but they usually surrounded grounds still dominated by formal geometric gardens. The ha-ha was not just an invisible barrier to keep intruders out of the garden and grounds. It was not simply the area that divided the stage upon which patricians played at paternalism safely separated from their hopefully awe-filled audience, the plebeians. The ha-ha was a device to make the gentry seem at one with their personal external environment, in which they could place themselves on top but within the safe confines of an invisible wall.[18]

Eighteenth-century English pleasure grounds were never truly natural. As M'Mahon explained, they were planned to look natural and were then decorated, often with classical Roman ruins or oriental ornaments.[19] Chinese garden concepts excited the European gentry, when first reports of them reached Europe in the eighteenth century; but Western attempts to emulate Oriental design usually resulted only in copies of their architectural features. Oriental architecture directly influenced some eighteenth-century Chesapeake homes and garden structures. Pagodas, garden houses with upswept eaves, and Chinese style bridges decorated a few Chesapeake grounds, but the garden spaces themselves were still divided into geometric partitions, often on terraced falls.[20]

Until recently, English garden historians generally agreed that by the end of the eighteenth century, very few formal gardens, with traditional geometric bed designs, remained in Britain.[21] Apparently, the British had reached a level of sophistication that allowed them the freedom to resist their highly structured civilization and their hedged landscapes. This change, combined with the need to conserve dwindling supplies of timber and game, led to the natural grounds movement in England. The movement did not spread quickly to America.

The end of the eighteenth century saw increased social stability in the colonies and the climax of a revolution in science, associated with Sir Isaac Newton, that resulted in fundamental changes in man's attitude toward the world about him. For the enlightened Chesapeake gardener, the garden nourished mind and spirit as well as body. The American pleasure garden became a visual expedient, combining the religious Eden myth with an evolving set of social and political goals, espoused by, among others, Thomas Jefferson and J. Hector St. John de Crèvecoeur. These religious and social concepts coincided with revolutionary new ideas about human beings' conceptual process that were shaped by John Locke and Joseph Addison.[22]

In the eighteenth century, Locke was interpreted to believe that visual images, such as those of the garden, were the primary conduit through which humans gained knowledge of external reality. Joseph Addison wrote of a spectrum of modes of perception, with the gross sensual pleasures at one pole and pure intellect at the other. The garden was an ideal illustration of Addison's conceptual theory, because it appealed to all the senses of the human animal, who tended to submerge these instincts as he became more cerebral. The goal was some balance of the two. Addison stated, "We find the works of nature still more pleasant, the more they resemble those of art." Nature and the garden were vehicles to sharpen both intellect and spirit. Just after reading Addison's works, one Chesapeake gentleman wrote to a friend, "The imagination acts intuitively; it seizes at once the sublimest parts as the eye catches objects. Nature, Hills, rocks, woods, precipices, water-falls rush upon the mind."[23]

Crèvecoeur saw the virgin American land filling Everyman's mind with irresistible aspirations, but he too believed that pure nature was not as inspiring as improved nature. Landscape should be ordered by humans, a collaboration of human vision and toil plus nature's spontaneous process. "This formerly rude soil has been converted by my father into a pleasant farm," he wrote, "and in return it has established all our rights." Crèvecoeur saw a direct relationship between ordering the land and gaining political freedom. He theorized that people, like plants, derived their "flavor" from the soil, and he declared that America's soil was still pure. Crèvecoeur believed that in America, with its newly emerging institutions, the relationship between people and the external environment they shaped around them was extremely important.[24]

Thomas Jefferson agreed with Crèvecoeur, and in his *Notes on Virginia*, Jefferson stated that the physical attributes of the land were less important than its metaphoric powers. The land was an image in the mind of the new American citizen, representing aesthetic, political, and religious values. In the eighteenth century, the garden was seen by many as an important visual determinant in the actions and responses of people. Even a clockmaker-innkeeper was aware of the impact of these ideas on his life in the newly emerging nation; among the names

William Faris gave the tulips he cultivated were "Sir Isaac Newton," "The Spectator," "Jefferson," and "The Farmer."[25]

Literate citizens of the new nation were looking to the Italian Renaissance and its classical antecedents for artistic and scientific knowledge, as well as for guidance in establishing their new republic. The 1783 catalogue of the circulating library in Annapolis and the 1796 catalogue of the Library Company of Baltimore offered their patrons Renaissance authors, such as Palladio, and their classical predecessors: Virgil, Horace, Pliny, and Columella.[26] Columella believed that agriculture and gardening were "sister to wisdom."

M'Mahon took advantage of this desire for scientific learning and zeal for the new nation. In the *American Gardener's Calendar* he asserted, "Producing a plentiful supply of wholesome vegetables and fruits, so necessary to health . . . as well as materials for ornamenting the whole face of the country . . . favour . . . acquisition of that useful branch of knowledge."[27] In fact, the garden had inspired and been a stage for displaying nationalism for years. Again the names William Faris chose for his tulips reflect his enthusiasms, for the new nation and for classical republican ideals. In his diary on July 3, 1801, Faris listed his tulip varieties by name. They included "General Washington," "Lady Washington," "General Montgomery," "General Warren," "General Mercer," "General Green," "General Williams," "General Wayne," "General Smallwood," "General Putnam," "General Harry Lee," "General Morgan," "General Gates," and "Colonel Howard." Faris also named his precious tulips after political leaders—"Adams," "Hamilton," "Madison," and "Dr. Franklin"—and for classical heroes—"Aristides," "Fabius," "Pompey the Grate," "Archimedes," "Cato," "Cicero," "Domostines," and "Cincinnatus."

Naming flowers after national and classical heroes was not peculiar to Faris. On April 9, 1804, he recorded in his diary receiving balsam plants from his neighbor Alexander Contee Hanson (1749–1806), whose father John (1721–83) had signed the Articles of Confederation and served as the first president of the Congress in 1781. The son was deeply affected by the Revolution and wrote, "during the whole memorable interval between the fall of the old and the institution of the new form of government, there appeared to exist among us such a fund of public virtue as has scarcely a parallel in the annals of the world."[28] It is not surprising that the younger Hanson named his balsam plants "General Washington," which Faris described as white, purple, and crimson; "Franklin," which was purple; "Lady Washington," "flesh mixed"; "the President," crimson and pink; "Aristides," white and purple; and "General Green," which Faris left undescribed. Hybridizing new varieties of flowers to be named for classical and national heroes became a popular pastime after the Revolution in the Chesapeake.

Even before the Revolution, colonists understood that flowers were inspira-

tional symbols for higher thoughts. In 1766 Charles Carroll of Carrollton wrote
to a friend, "If you have a turn for gardening or for exotick Plants & flowers I shall
perhaps be able to send you such of these which as uncommon in England may
afford you some pleasure as a florist, or matter of thought & speculation as a naturalist, or Philosopher."[29] Flowers could signify a personal friend as well as a distant hero. William Gordon wrote George Washington in 1786, "Shall I endeavor
to furnish your garden . . . with flowers and plants that may keep up the remembrance of an absent friend."[30]

M'Mahon did not name his flowers after friends or national heros. He preferred to use the new scientific classifications being developed in Europe. He used
what he called "classical catalogue" definitions of all plants.[31] In 1808, when
M'Mahon decided to expand his business and create a botanical garden near
Philadelphia, he named the garden Upsal after the Swedish university where bot-
anist Carl von Linne developed the Latin binomial system of plant classification
and nomenclature. In his gardening treatise, M'Mahon encouraged his readers to
become scientists, as the Italian Renaissance model promoted.

Well before M'Mahon, gentry and middling sorts alike in the colonial Chesapeake were fascinated by the scientific aspects of gardening. Like elite gardeners
Washington and Jefferson, craftsman Faris, as his diary shows, was a careful chron-
icler of nature. He experimented with growing various plants, especially tulips; he
grafted and selectively cross-pollinated. He consistently recorded his observations
of the weather and its effect on his garden. He and his contemporaries used their
gardens to understand the order of nature and to subject it to their own order in
terms of design, plantings, and processes. As M'Mahon wrote, gardening allowed
the gentleman to project his abstract ideas into nature.[32] Following the concepts
of optics from the Italian Renaissance, gardeners manipulated space, by using line
and proportion to create a favorable and safe perspective of the world around
them as viewed from within their gardens and favorable perspectives of themselves when their gardens were viewed from without.

During the latter half of the eighteenth century, the English landed gentry became enamored of the idea of scientific "improvement" and began to retire to
the country to run farms themselves. An English friend wrote to Charles Carroll
of Annapolis advising that he encourage his son into "bettering or ornamenting
his estate, improving his roads, erecting houses or beautifying his own habitation
in building walks, gardens, plantations, pleasure grounds, etc." And indeed, Chesapeake gentry and the middle classes soon took to improving their properties. They
began to drain and enclose and plant for ornament as well as use. During this
period, a Chesapeake gentleman's garden became a measure of his gentility just
as it had been for the educated man of the Italian Renaissance. In spite of their
beliefs about the evils of excess, many of the gentry of the eighteenth-century

Chesapeake were eager to reflect European taste in costume, entertainment, and housing. According to travelers, the Chesapeake soon boasted one of the highest societies on the Atlantic. A French hairdresser was on the scene in Annapolis, and the town's women were touted as some of the most genteel in the colonies. The number of artisans and craftsmen in the region's towns grew as the appetite for stylish luxury goods increased and craftsmen such as Faris became needed to supply the desires of the socially aspiring.[33]

The high cost of gentility in the Chesapeake excluded many of the middle classes from the stylish affairs of the bon ton, but the garden became one aspect of gentility that could be achieved by most classes in the emerging republic, with attention to discipline rather than acquisition of indulgences. After all, plants multiplied; fashionable goods and services were consumed. When cultivated into a garden, land became an area of common ground between the upper and middling classes, a place where genteel civility as well as plants could be cultivated and shared; and some of the fruits of such collaboration could even be eaten.

From William Faris's diary we learn that the elite and the common man were discussing, trading, and growing edible and ornamental plants. The relationships between rich and poor perpetuated by gardening endeavors confused English visitors to Maryland, both before and after the Revolution. In a letter back to England in 1772 William Eddis wrote, "An idea of equality also seems generally to prevail, and the inferior order of people pay but little external respect to those who occupy superior stations." Almost thirty years later Richard Parkinson wrote, "Now, with regard to the liberty and equality . . . among the white men in America, they are all *Mr.* and *Sir* so that in conversation you cannot discover which is the master or which is the man."[34]

Gardening was an area of commonality across the social strata of the new nation. It was not taking tea or dancing together; it was more basic, more unifying, even spiritual. The garden produced physical sustenance and inspiring order and beauty, and it elevated all parties to a more virtuous plane, where differences of class blurred. The garden was the space between nature and culture where each man could negotiate his individual position. Here one could order a small corner of the world and each spring begin again. In 1784 Nancy Shippen, a Philadelphian whose philandering husband had spurned her and assumed custody of their only child, retreated with her mother to a country house that was "pleasantly situated on a hill with a green Meadow before it." Behind the house were "a garden and a nursery of trees," to which she directed her attention. She wrote in her journal of the consolation she expected to find there. Although she could not help feeling like an outcast, "with all these conveniencys," she declared, "I ought to be contented."[35]

For centuries gardening had appealed to some fundamental spiritual need of humans, whose religions traditionally depicted a garden as the ideal abode for mankind on this earth and beyond. The ordered garden was, after all, Everyman's refuge from the terrifying unknown, the evil wilderness. The garden offered sanctuary from the threat of wild nature and escape from barbarian outsiders. The great garden of the vast American frontier held some frightening connotations for most colonists. Michael Wigglesworth wrote of it in 1662,

> *A waste and howling wilderness,*
> *where none inhabited*
> *But hellish fiends, and brutish men*
> *That devils worshiped.*[36]

America was viewed by some as a seedbed in which to establish natural spirituality, and gardening was one method to nurture higher values. John Beale Bordley gave up the public life in Annapolis to pursue experimental agriculture and moved to a 1600-acre Wye Island estate he acquired in 1770. He was instrumental in founding, in 1785, the Philadelphia Society for Promoting Agriculture, an association whose membership included twenty-three Marylanders by 1798. In his book *Essays and Notes on Husbandry and Rural Affairs*, published in Philadelphia in 1797, Bordley offered his ideas on keeping workers happy on the farm. He suggested that each worker be given a garden 80, 90, or 100 feet square, because "it was observed by a clergyman . . . cottagers who had a garden were generally sober, industrious and healthy; and those who had no garden, were often drunken, lazy, vicious and ailing." The garden was a symbolic religious battleground, where good battled evil, where temptation and sin were overcome by forgiveness and reconciliation. Here too, dangerous "intemperance" could be rectified, according to M'Mahon. It is interesting to note that there is a high correlation between those with whom William Faris shared church membership and those with whom he exchanged plants and gardening advice. Even though it was twenty years after the colonial period of mandatory church attendance, the people Faris came to know through nearby St. Anne's Church formed the nucleus of his pleasure gardening colleagues.[37]

The evils of avarice and the injustices of power politics drove even wealthy colonists to seek spiritual refuge in a nature that they ordered around themselves. In 1771, as frustrations with England mounted, a future signer of the Declaration of Independence, Charles Carroll of Carrollton, wrote to a friend, "The wisest Philosophers, the greatest poets, and the best men have constantly placed the most perfect sublime happiness in rural retirement. Under the shades of Forrests statesmen have sought happiness having in vain sought after it in the perplexed mazes of ambition and interest."[38]

Bernard M'Mahon, like his friend Jefferson, well understood that nature in general and particularly gardening—the ordering of nature—were intertwined with morality and nationhood in the minds of the political leaders as they structured the fledgling nation's emerging institutions. In 1779, when Jefferson was governor of Virginia and a member of the Board of Visitors at the College of William and Mary, curriculum reforms resulted in the appointment of Robert Andrews as "Professor of Moral Philosophy, the Laws of Nature and of Nations, and of the Fine Arts." Jefferson defined the fine arts as "Sculpture, Painting, Gardening, Music, Architecture, Poetry, Oratory, Criticism." In the 1783 *Catalogue of the Annapolis Circulating Library*, where books were grouped in categories, the section containing books on pleasure gardening was titled "Gardening, Poems, Plays, etc."[39] Jefferson even wrote of his garden in terms of art. In 1807 he wrote, "The canvas at large of a Garden or pleasure grounds must be a Grove, of the largest trees . . . trimmed very high, so as to give the appearance of open ground."[40]

Even though nurseryman M'Mahon later promoted gardening to every segment of society in the new nation, it was evident that before the Revolution true pleasure gardening as a "fine art" was only theoretically accessible to every man in the emerging republic. All the aspiring "artist" needed was an excess of land and leisure time; some knowledge of the rules of perspective, classical design, mythological symbolism, and horticulture; regularly available labor not otherwise needed to produce income; and the inclination to present himself at the pinnacle of the hierarchy of nature as he had ordered it. Gardening primarily for ornament in the eighteenth century was obviously limited. Even after independence, the pleasure gardener of the emerging republic was primarily the property owner, the male citizen of the United States of America.

In the Maryland landscape paintings of Francis Guy, it was usually the male owner, often accompanied by a male visitor, who was depicted surveying his ornamental grounds. The pleasure-gardening property-owning male in the Chesapeake was usually also a slave owner and often rented others' slaves or paid free blacks to help shape and maintain his personal external environment. The possession of capital was an important ingredient in determining who pleasure gardened.

In most Chesapeake homes, women might be in charge of the greenhouse or the everyday activities in the kitchen garden, but they were not often the master of the grounds. This was not always true. Eliza Lucas Pinckney, a strong-willed and rich gentlewoman of South Carolina, wrote in a 1742 letter to a friend of the garden she was planning: "it shall be filled with all kind of flowers, as well wild as Garden flowers, with seats of Camomile and here and there a fruit tree—oranges, nectrons, Plumbs." As the nineteenth century dawned, women began to play a more important role in planning the garden. At the Riversdale plantation in Prince George's County, Maryland, Rosalie Stier Calvert wrote to her father on May 19, 1805, "We are getting much better at the art of gardening."[41]

Women sometimes referred to their gardens as amusements. Rosalie Calvert, in another letter to her father, wrote, "I want to make the garden my principal amusement this summer," and Eliza Pinckney in 1760 wrote, "I love a garden and a book; and they are all my amusement." The gardens at the Girls' School in Salem, North Carolina, were described as "designed for literary repast, & evening amusement."[42]

Men were not above simply amusing themselves in their gardens either. George Washington reported that gardening had become his amusement. The Reverend William Bentley, describing the garden of Boston merchant Joseph Barrell, wrote, "Was politely received by Mr. Barrell who shewed me his large and elegant arrangements for amusement and philosophical experiment." Joseph Barrell's garden was his stage. Here he excitedly explained each garden plant and feature to his exhausted guest until well after dark.[43]

In fact, M'Mahon referred to garden terraces as theatrical arrangements,[44] and the 1783 Annapolis book catalogue grouped gardening and plays together. Gentlemen of the Italian Renaissance used their gardens for theatrical presentations. Townspeople up and down the Chesapeake were very familiar with devices of the theater. Plays had been performed in Williamsburg for years, and a playhouse opened in Annapolis in 1752 next door to William Faris's home and shop.[45] And of course, although an eighteenth-century gentleman's garden might never be used for a theatrical presentation, it was the outdoor platform he designed and on which he presented himself to his visitors and to the community at large. Manipulating the view for the sake of the visitor was a continuing theme throughout M'Mahon's treatise. The great and the not so great enjoyed garden watching.[46]

The view of country seats and gardens sitting high up on the American landscape inspired patriotic feelings in some observers. Of Joseph Barrell's grounds one wrote in 1794,

> *Where once the breastwork mark'd the scenes of blood,*
> *While Freedom's sons inclosed the haughty foe,*
> *Rearing its head majestic from afar*
> *The venerable seat of Barrell stands*
> *Like some strong English Castle.*[47]

American forefathers like George Washington, Thomas Jefferson, and John Adams were aware of directions in garden design that had been spearheaded by the political leaders of centuries past and which were the basis for gardens in the British American colonies of the eighteenth century. The spiritual aura and popular importance of gardens was not lost on these men. George Washington, who devoted much time and effort to organizing his garden, wrote that gardening was among "the most rational avocations of life." He believed, as did Bordley, that gardening contributed to the spiritual health as well as the amusement. Jefferson wrote, "no occupation is so delightful to me as the culture of the earth, and no culture comparable to that of the garden." But he saw the age of the garden ending as he wrote of the British during the War of 1812, "Our enemy has indeed the consolation of Satan on removing our first parents from Paradise: from a peaceable and agricultural nation, he makes us a military and manufacturing one."[48]

The garden did wither as a symbol of power and moral force as the agricultural gave way to the industrial and factories flowered on the American landscape. As the gentility of pleasure gardening became available to greater numbers of the middling sort in the emerging republic, it declined in importance to the ruling class, and the symbol of might and right shifted from the garden to the machine. William Faris was a clockmaker at the end of the America's agricultural age, when people's perception of time still relied on nature's manifestations, the rising and the setting of the sun and the changing of the seasons. Industrialization would dramatically change the significance of time and the clock.

The clock would soon become the mechanical indication of units by which work, and therefore pay and worth, were measured. In the agricultural economy of eighteenth-century America, a man's worth was measured by his harvest; and he made a pleasure garden to control, in an abstract and artful form, at least a small part of unpredictable nature, which otherwise controlled him. The setting sun halted his day's work. An unexpected storm or drought could destroy his daily plans or his yearly harvest.

Bernard M'Mahon and William Faris were wedged between the old agricultural order and the coming technical age. The looming industrial era would see cities burgeon and replace the wilderness as the frightening place in the minds of the American people. Citizens of the machine age would retreat to bucolic woodlands for the serene security of nature, much as citizens of Faris's era clamored for the safety of towns with ordered streets and tidy gardens when threatened by the terrifying unknown lurking in the uncivilized nature of the frontier.

Gardening would become just one more diversion among many in an industrial and technological world where individuals' livelihoods were no longer dependent on manipulation of the land. M'Mahon and his colleagues contributed to this trend in nineteenth-century America, as they promoted gardening to all classes and both sexes in the new nation. M'Mahon hoped his book might "make any person . . . his own Gardener."[49]

Becoming a gardener helped a person understand the cycle of life and death, and many Chesapeake gardeners chose to bury their loved ones in their gardens and went there to remember departed relatives and friends. If the spiritual garden was the place we all began, they reasoned, then it was comforting to return to the garden when we died.[50] Where sufficient land was available, a cemetery was often created adjacent to the garden. As one traveler recorded in 1790, "It is very common to see in large plantations in Virginia, and not far from the dwelling house, cemeteries walled in, where the people of the family are all buried. These cemeteries are generally built adjoining the garden."[51] Employees as well as relatives were buried in Chesapeake plantation gardens. At Nomini Hall on June 23, 1789, Robert Carter recorded, "On Saturday the 20th June Mr. George Randell departed this Life & his Remains were interred in the Garden near to the Grave of Mr. Jos. Taylor School Master."[52]

Another reason early Americans might have preferred burial in their own gardens was security. In his journal on January 29, 1774, Philip Fithian, visiting Nomini Hall in Virginia, quoted his host, Robert Carter, on this subject: "he much dislikes the common method of making Burying Yards round Churches . . . almost open to every Beast. . . . he would choose to be laid under a shady Tree where he might be undisturbed, & sleep in peace & obscurity—He told us, that with his own hands he planted, & is with great diligence raising a Catalpa-Tree at the Head of his Father who lies in his Garden." In fact, some felt that burying the dead in a common community or church cemetery made the sight and thought of death too familiar. One observer commented, "Instead of producing those solemn thoughts and encouraging those moral propensities . . . it renders death and the grave such familiar objects to the eye as to prevent them from awakening any serious regard . . . and . . . to eradicate every emotion naturally excited by the remembrance of the deceased."[53]

Early Americans realized that gardens, economies, and men are ultimately under nature's control. Perhaps they found some comfort in the knowledge that nature, not man, renews life year after year. In an agrarian society, people understood the symbiotic relationship of people, plants, soil, weather, and the seasons. They understood that there is an order to nature but not always a kindness. People whose lives depended on the success of crops understood that nature controls floods, hailstorms, droughts, tornadoes, and death.

Eighteenth-Century Chesapeake Gardens Today

J ust as they did in the eighteenth century, Americans today love to visit gardens. The historic Virginia gardens of George Washington and Thomas Jefferson are maintained with historical integrity and are surely worth a visit.

Washington's Mount Vernon is just a short drive from the District of Columbia on a parkway that winds south along the scenic Potomac River's path toward the Chesapeake Bay. Driving along the Potomac, you can imagine the beauty the area possessed in the eighteenth century and can realize why Washington was obsessed with having the nation's capital there. At Mount Vernon the tourist can still see the fleur-de-lis-shaped beds that Benjamin Latrobe bemoaned more than two hundred years ago. Even today, some may find such traditional geometric walled garden spaces static and uninteresting, if we think of a garden as a process rather than a product, we realize that Washington's garden changes with each new day's growth. Washington's favorite plantings grow up and grow together adding depth and energy to a space that is defined by paths, walls, and buildings arranged to be visually satisfying even in the winter, when the plants die back.

Washington's garden schoolhouse remains standing in one corner of the flower garden. The greenhouse, to which Margaret Carroll, of Mount Clare in Baltimore, insistently donated her favorite lemon and orange trees, still houses such plants, which are painstakingly moved from the greenhouse to the garden for each summer season (see Plates 20 and 21). Washington, a man very concerned about retaining an appearance of austere dignity, might blush to think that many thousands of people now know that he toured his grounds each morning with his beloved dogs, True Love, Sweet Lips, and Mopsy.[1] Recently, the Mount Vernon Ladies Association, which owns and operates the site, began offering for sale in their shop a variety of authenticated reproductions of garden furniture and ornaments that Washington himself might have used.

Further to the south, and landlocked far from the Chesapeake Bay, Thomas Jefferson's homesite, Monticello, sits high atop a small mountain overlooking his beloved University of Virginia in Charlottesville. Here geometry defines the vegetable and fruit gardens but runs amok in the flower gardens that border the serpentine path encircling the lawn. Flowers that Jefferson loved merge and play against one another in unexpected and dramatic profusion. Jefferson's carefully orchestrated backdrop of trees completes the canvas one views, looking out over the lawn and flowers. Here Jefferson's abstract vision grows into towering reality.

Monticello boasts one of the few fishponds to survive from the eighteenth century. It sits next to the garden path, close enough to house and kitchen that you can almost smell the fish sizzling in the dinner pan. Jefferson played the angularity of his buildings and practical gardens beds against the curves of his serpentine flower gardens, his trees and groves, and the natural mountain, to add an unexpected complexity to the experience, a balance of control and freedom.

At the Thomas Jefferson Center for Historic Plants, established in 1987, the visitor may buy historic plant varieties that Monticello gardeners collect and preserve on an adjacent farm, which, like Monticello, is owned by the Thomas Jefferson Memorial Foundation. Seeds of plants grown by both President Thomas Jefferson and craftsman William Faris are available, including cockscomb (*Celosia cristata*), native columbine (*Aquilegia canadensis*), hollyhocks (*Alcea rosea*), and Job's tears (*Coiz lacryma-jobi*). But the most appealing by far is a delicious-smelling native shrub, adored by gentry and their lessers alike, the heavy-scented shrub *Claycanthis floidus*, with its deep maroon flowers that smell sweeter and more seductive than dead-ripe strawberries on a warm spring night. Jefferson was so taken with this plant that he shipped samples of it to friends in Paris. The center even carries reproductions of flower pots archaeologists unearthed at Monticello, as well as of the pot Rembrandt Peale included in his famous 1801 portrait of his bespectacled brother Rubens gently caressing a favorite geranium.

Traveling eastward through Virginia, back toward the Atlantic, we find 90 acres of America's most famous reproduction of small town gardens, at Colonial Williamsburg. Often criticized as high-end colonial revival fantasies, most of these ninety-nine gardens were designed in the 1930s, 1940s, and 1950s by Williamsburg's preeminent revivalist, landscape architect Arthur Shurcliff. While they resemble in design the planned geometric garden of Annapolis's William Faris, their plantings tend less often to be edible than those one would have found in Faris's very economically laid out garden. We are lucky to have the gardens at Williamsburg; Faris's is currently covered by the concrete of a bank parking lot.

Sensitive to the whispers and titters about colonial revivalism, the archaeological, research, and landscape professionals at the Colonial Williamsburg Foundation began, in 1994, to reexamine the grounds, letters, and other documents of circuit court judge St. George Tucker, who bought his house in Williamsburg in 1788. When restored in 1931, the Tucker garden contained the requisite boxwoods, peonies, bowling green, and garden house, the staples of all worthy colonial revival gardens.

The research proved that Tucker was one of those dyed-in-the-wool pragmatic gardeners of the new republic who was interested in both design and in practical cultivation. His particular interest was fruits and nuts. Archaeology revealed that his garden had the predictable broad central walkway with two cross paths, plus what may have been a line of fruit trees on the left side of the property, while the right side was defined by a fence. Apparently Tucker also planted a collection of his favorite "sweet and ornamental shrubs" close to his house, where soft breezes could fill his rooms with pleasant lingering aromas throughout the growing season.[2]

The good news about the urban gardens scattered throughout the restored section of Williamsburg is that a visitor can walk the streets and see most of them without paying a fee, with the notable exception of the early terraces at the Governor's Palace. Terraces from about 1747 grace the garden of the Robert Carter house, next to the palace. Recently, the Colonial Williamsburg Foundation has designated a special section of the historic area as the Colonial Garden, in which their best gardeners demonstrate period tools and techniques and even sell samples of some colonial plants. The main reception center's gift shop usually carries a fine selection of books on eighteenth-century gardening and crafts.

The garden tourist should also plan to drive to a nearby Colonial Williamsburg property, Carter's Grove, built by Lewis Burwell in 1751, where a restored terraced garden falls in gentle turfed flats toward a large fenced rectangular garden and eventually to the James River. The house is a colonial revival restoration; and the fenced garden, although a little overdone here and there, is huge and impressive nonetheless. Only a few noisy helicopters and annoying unplanned trees mar the view from the house to the river. An unexpected plus is a rare opportunity to view reconstructed slave quarters, with their swept yards and small huck patches.[3]

Searching for those traditional falling terrace gardens that helped define the eighteenth-century Chesapeake is a joyous adventure. Traveling north from Williamsburg, the garden hunter can visit Oatlands at Leesburg, Virginia. Built in 1804, this falls garden features boxwood and enough springtime bulbs to thrill even William Faris.

Privately owned Sabine Hall, built by Landon Carter in the mid-eighteenth century on the north side of the Rappahannock River in Richmond County, Virginia, also possesses grass-ramped terraces falling toward the river. This design was a youthful Carter's respectful nod to the Italian Renaissance. In later years he wrote "as an old man . . . my Colic will not let me, as I used to do, walk out and injoy the pleasure of flowers." And since he could not stroll in his garden anymore, he planted the entire area in turnips in 1777.[4] Future owners banished the turnips and recalled the flowers from their temporary exile. Terraced gardens from the eighteenth century or the traces of them dot the Virginia countryside.[5]

Closer to the Chesapeake Bay, in Annapolis, it is easy to imagine the practical fruits and vegetables that filled the geometric squares in the 1770s terraced garden of the very wealthy Charles Carrolls, father and son, as you visit their home on Spa Creek. The octagonal brick garden houses that sat at either end of the 400-foot creekside walkway are long gone, but some of the brick wall remains, as do the distinct terraces leading down to the water from the house. Unfortunately, the garden plantings themselves no longer reflect the eighteenth century, but the view of sailboats passing by on Spa Creek makes up for that disappointment.[6]

A few blocks away, the visitor can tour the 1760s garden of Governor William Paca, which was resurrected from under a hotel parking lot in the 1960s. It, too, is a terraced falls reconstructed to emulate generic-but-pleasing decorative garden squares on the flats. This was the best that could be accomplished by way of restoration, since few Paca records about the garden's actual plantings exist. Archaeological studies accurately determined the angles of the falls and unearthed the surrounding brick walls. The bottom portion of the garden, a surprising naturalized area, is more nearly correct, because it appears in a portrait of Paca painted by Charles Willson Peale (see Plates 14, 15, and 16). Gardeners and horticulturalists employed by Historic Annapolis, Incorporated, take great care to document plants from the period, and they hold heirloom plant sales in the spring. Leaving Annapolis without a nod in the direction of the Maryland State Archives would be unconscionable, for here are people and sources enough to inspire any colonial historian to trudge onward.

Not far from Annapolis, in Prince George's County, Maryland, is Belair Mansion, begun in 1745 as a country retreat for Samuel Ogle, provincial governor in the 1750s and '60s. The entrance to the estate shows the remains of a mile-long alley of tulip poplar trees, and the garden façade has three terraces connected by grass ramps falling away from the house. At the bottom of the garden is the family cemetery. The grounds are owned and maintained by the City of Bowie, which is determined to learn as much about its garden treasure as possible.

In Baltimore, the garden explorer can visit two falling gardens. Near downtown is the Mount Clare estate, where the entrance gates, walls, statues, and courtyard have been restored according to evidence in contemporary paintings, and where extensive archaeological studies are helping historians restore the terrace garden described by Mary Ambler in 1766. The original three flats were cut into five levels sometime in the nineteenth century, but the board of the Carroll Park Foundation, which controls the land, is determined to restore it to its colonial glory. The adjoining orchard and grape arbor are also under cultivation and growing even as you read this.

North of Baltimore City is Hampton Mansion, run by the National Park Service. Here the terraced falls connected by steep grass ramps have not been disturbed. Although the plantings on the flats reflect a later, nineteenth century interpretation, the ramps are spectacular because of their sharp-angled, lush green authenticity.

Remnants, restorations, and reconstructions of traditional Chesapeake gardens can be found throughout Maryland and Virginia. With the help of local historical and garden societies and of state and county preservation offices, the garden explorer can discover this wealth. Happy hunting!

Appendix

The diary and account books of William Faris provide an extraordinary record of the gardening practices of a middle-class resident of the Chesapeake at the end of the eighteenth century. The information in the following tables offers a detailed look at what he grew in his own garden and what additional produce he and the guests at his inn consumed.

Table 1.　Plants Grown by William Farris, in Annapolis, 1792–1804

PLANT	DATE FIRST MENTIONED IN DIARY
Shrubs, Trees, Vines, Grasses	
Althea	03/22/98
Boxwood	03/26/92
Holly Tree	04/01/97
Horse Chestnut	01/15/93
Ivy	03/26/93
Formoso	03/15/04
Pride of China Tree	04/06/96
Ribbon Grass	03/31/92
Snowball	08/13/92
Strawberry Tree	04/01/94
Sweet Scented Shrub	03/18/93
Tallow Tree	04/07/96
Willow Tree (Golding)	03/26/93
Flowers	
Anemonie	05/14/93
Asters	04/07/92
Balsam Apple	03/29/93
Bleeding Heart	08/13/97
Callamus	10/05/03
Carnation	05/05/92
Chrysanthemum	04/21/99
Columbine	04/07/92
Coxcomb	04/07/92
Crocus	04/18/03
Crusula	04/26/98
Daffodil	05/01/94
Emmy	04/26/98
Flowering Pea	04/03/94
Fluer de Leis	01/17/93
Geranium	05/05/01
Globe Amaranthus	04/22/99
Hollyhocks	08/10/01
Hyacynth	03/28/94
Iceplant	03/06/93
Impatiens	04/07/92
India Pink	04/07/92
Iris	08/11/01
Jacobson Lily	03/18/02
Jasmine	01/17/93
Jerusalem Cherry	02/03/93
Job's Tears	04/02/94
Jonquil	05/01/94
Lady In Green	04/08/93

(continued)

PLANT	DATE FIRST MENTIONED IN DIARY
Lily	05/26/98
Lily of the Valley	04/03/94
Marigold	04/07/92
Migonette	05/15/92
Narcissus	05/01/94
Nasturtium	04/18/01
Parson's Pride	04/19/98
Passion Flower	01/17/93
Polyanthus	03/06/92
Poppies	04/07/92
Primrose	04/07/92
Reason	02/03/03
Rose	04/07/92
Satin Flower	04/03/94
Sensitive Plant	04/07/92
Snapdragon	05/02/95
Ten Commandments	05/19/04
Tube Rose	04/07/92
Tulips	03/24/92
Virginia	04/07/92

Vegetables	
Asparagus	03/20/92
Beans	04/03/92
Beets	03/29/92
Broccoli	05/25/97
Brussel Sprouts	05/25/95
Cabbage	03/15/92
Cantaloupe	03/17/94
Carrots	03/21/92
Cauliflower	04/08/93
Colewart or Kale	03/28/92
Corn	04/10/97
Cucumbers	03/17/94
English Lambs Quarter (Orach)	04/08/93
Egg Plant	03/06/93
Garlic	04/06/96
Greens	03/31/92
Leeks	08/22/93
Lettuce	03/23/92
Mush Mellons	03/23/92
Nasturtium	04/18/01
Onions	05/04/92
Parsnip	03/19/92
Peas	03/23/92
Pepper	05/04/99
Pumpkin	04/17/04

(continued)

PLANT	DATE FIRST MENTIONED IN DIARY
Radish	03/19/92
Savory	05/26/98
Shallots	03/24/95
Squash (Simlins)	03/02/95
Spinach	03/21/92
Turnip	05/25/97
Watermelon	03/24/92

Herbs	
Bergamot Balm	08/25/94
Catnip	05/13/03
Ginger	09/01/02
Horseradish	09/17/99
Mint	07/15/02
Nutmeg (Indian)	04/19/97
Parsley	03/27/92
Pickling Lime	04/27/92
Poppy	04/07/92
Rosemary	11/26/93
Saffron	10/09/98
Sage	03/26/92
Thyme	04/14/95

Fruits and Nuts	
Apple	08/04/95
Almond	09/03/02
Cherry	03/25/93
Currant	03/02/95
Gooseberry	03/18/93
Grapevines	03/07/96
Mulberry	05/25/95
Peach	03/17/94
Pear	03/17/99
Plum	09/05/99
Walnut	04/03/92

Source: William Faris's Diary, 1792–1804, MS 2160, Maryland Historical Society (original spellings).

Table 2.   Produce Purchased by William Faris, in Annapolis, 1790–1804

	DATE IN ACCOUNTS	QUANTITY	PRICE (pounds/shillings/pence)
		Vegetables	
Beans	.09/26/95	½ bu	£2/0
Cabbage	.03/10/91	20 cabbage	£2/1
Cantalopes	.08/16/94	1 dozen	£0/6
Cauliflower	.03/24/91	1 barrel	£1/196 + fr.
	09/10/92	1 barrel	£1/13 + fr. £1/0
Corn	.01/30/90	2½ bu.	
	02/09/90	1½ bu.	£4/0
	03/03/90	6 bu.	£3/9 per bu.
	01/07/91	2½ bu.	£7/6
	03/05/91	2 bu.	£5/6
	07/18/91	1 barrel	£15/0
	07/20/91	4 bu.	
	06/29/92	6 bu.	£12/6
	09/19/92	3 bu.	£3/9 per bu.
	01/24/93	5½ bu.	£3/0 per bu.
	06/22/93	1 bu.	£3/0 per bu.
	08/28/93	1½ bu.	£4/1-1/2
	10/22/93	1½ bu.	£4/0 per bu.
	10/24/93	1½ bu.	£3/9 per bu.
	10/10/95	1½ bu.	£5/6
	10/16/95	2 bu., ½ pk.	£4/6 per bu.
	10/26/99	2½ bu.	£11/3
	12/24/99	1½ bu.	£6/0
Cucumbers	.07/14/90	1 doz.	£0/6
	07/24/90	6	£0/3
	08/01/92	1 doz.	£0/3
	07/11/93	1 doz.	£0/5
	07/13/93	1 doz.	£0/4-1/2
	07/16/93	1 doz.	£0/6
	07/19/93	1 doz.	£0/5
	07/23/93	1 doz.	£0/4
	07/25/93	½ doz.	£0/1-1/2
	07/03/94	1 doz.	£0/9
	07/16/94	½ doz.	£0/2
	08/16/94	1 doz.	£0/6
	07/21/95	1 doz.	£0/4
	08/04/96	1 doz.	£0/3
	07/01/97	15 cuc.	£1/4
Greens	.12/23/97	4 heads	£0/1

(continued)

	DATE IN ACCOUNTS	QUANTITY	PRICE (pounds/shillings/pence)
Hominey Beans	11/24/91	1 pk.	£1/2
Irish Potatoes	11/12/91	12 bu.	£18/0
	09/03/97	½ bu.	£2/6
	10/16/98	1 bu.	£3/9
	09/02/00	1 pk.	£1/6
Mellons	08/16/90	4 mellons	£1/9
	09/02/97	1 mellon	£0/5
Mushmellons	08/16/90		
	08/16/91	1	£0/3
Onions	09/03/97	½ bu.	£3/9
Peas	06/03/97	1 pk.	£1/11
	05/19/98	1½ pk.	£3/4
Potatoes	03/15/90	1 bu.	
	10/10/90	1 bu.	£1/6
	10/31/91	1 bu.	£3/0
	11/25/91	10 bu.	£15/0
	10/17/93	1½ bu.	£4/0
	10/26/93	½ bu.	£1/0
	03/08/94	2 bu	£2/0 per bu.
	03/16/94	⅓ pk.	£0/11
	09/26/94	3 pk.	£1/6
	10/29/94	1 bu.	£3/0
	03/17/95	1 pk.	£1/6
	12/08/97	3½ bu.	£13/1-1/2
	12/10/97	2 bu., 3 pk.	£10/3-1/2
	10/01/98	½ bu., ½ pk.	£3/1
	10/13/98	1½ bu.	£7/6
Rice	03/14/90	3 lb.	£1/0
	04/10/90	3 lb.	£1/0
	06/16/90	3 lb.	£1/0
	12/27/92	6 lb.	£2/0
	12/22/97	6 lb.	£2/9-12
	02/02/98	6 lb.	£2/6
	03/14/98	5 lb.	£2/4
	04/23/98	6 lb.	£2/6
	11/17/98	3 lb	
	11/07/99	1 lb.	£5/0
	03/03/00	2 lb.	£2/11
	08/16/00	1 lb.	£0/0/8

(continued)

	DATE IN ACCOUNTS	QUANTITY	PRICE (pounds/shillings/pence)
Simlins (Squash)	07/28/91	14	£0/7
	07/29/91	1 doz.	£0/6
	07/30/91	1 doz.	£0/4
		1 doz.	£0/6
	08/08/91	1 doz.	£0/6
	08/13/91	1 doz.	£0/7
	07/03/93	1 doz.	£0/7
	08/05/93	1 doz.	£0/6
	08/12/93	1 doz.	£0/6
	08/24/93	1 doz.	£0/5
	07/16/94	1 doz.	£0/6
	07/25/97	20 simlins	£0/3
	07/29/97	½ doz.	£0/3
	08/04/97	1 doz.	£0/4
	08/08/97	1 doz.	£0/4
	08/12/97	1 doz.	£0/5
	08/19/97	1 doz.	£0/5-1/2
	08/11/98	1½ doz.	£0/7
	08/18/98	1 doz.	£0/4
	08/25/98	1 doz.	£0/4
	08/29/98	1½ doz.	£0/5-1/2
Sweet Potatoes	10/23/92	1 bu.	£3/0
	11/01/92	6 bu.	
	09/29/95	2 bu.	£7/6
	10/07/97	½ bu.	£2/6
	11/04/97	3 pk.	£3/0
	09/22/98	bushel	£3/0

Fruits and Nuts

	DATE IN ACCOUNTS	QUANTITY	PRICE (pounds/shillings/pence)
Apples	09/14/91	10 doz.	£4/6
	12/20/92	3 doz.	£1/10
	08/02/00	1 doz.	£0/2
Chesnuts	10/12/98		£0/5-1/2
	10/14/98		£1/4-1/2
	10/29/99	9 qts.	£3/5
Currants	12/20/91	5 lb.	£5/0
	12/12/92	5 lb.	£6/8
	12/22/94	4 lb.	£6/0
	12/23/95	5 lb.	£5/0
	12/13/96	4 lb.	£7/6
	12/17/98	5 lb.	£7/6
Huckleberries	07/05/00		£0/5

(continued)

	DATE IN ACCOUNTS	QUANTITY	PRICE (pounds/shillings/pence)
Lemons ....06/29/99 ....1 doz. £3/9			
Limes ....10/22/99 ....1 bu. £2/6			
	08/06/00 ....2½ bu. £4/2		
	08/07/00 ....1½ bu. £2/6		
	08/09/00 ....½ bu. £0/10		
Pears ....09/17/99 ....£1/5			
Plums ....12/23/95 ....2 lb.			
	12/17/98 ....6 lb. £7/6		
	12/31/99 ....2 lb. £4/0		
Raisins ....12/20/91 ....4 lb. £4/0			
	12/12/92 ....4 lb. £5/0		
	12/20/94 ....4 lb. £4/0		
	12/23/96 ....2 lb. £3/9		
	12/16/99 ....6 lb. £12/0		
	12/23/99 ....1 lb. £3/0		
Strawberries ....06/03/98 ....£2/7-1/2			
	05/27/00 ...."Strawberries and Milk" £0/9		

Source: William Faris Account Books, MS 1104, Maryland Historical Society (original spellings).

Notes

Preface

1. The most enlightening recent book on gardening in eighteenth-century England is Tom Williamson, *Polite Landscapes: Gardens and Society in Eighteenth-Century England* (Baltimore: Johns Hopkins University Press, 1995).

2. One of the best explanations of how the concept of rural retirement was realized in the British American colonies may be found in C. Allan Brown's article "Eighteenth-Century Virginia Plantation Gardens: Translating an Ancient Idyll," which appears in *Regional Garden Design in the United States*, ed. Therese O'Malley and Marc Treib (Washington, D.C.: Dumbarton Oaks Research Library and Collection, 1992).

3. M. L. E. Moreau de St. Mery, *Moreau de St. Mery's American Journey, 1793–1798*, ed. and trans. Kenneth Roberts and Anna Roberts (Garden City, N.Y.: Doubleday, 1947), 121. Some contemporary European visitors heaped lavish praise on the American "villas" dotting Virginia's rivers. Allan Brown speculates that their comments might have exemplified "their capacity to see beyond the actual to the signified" (O'Malley and Treib, *Regional Garden Design*, 139).

4. In England, hunting had long been a mark of social status. Legislation ensured that deer and other forms of game, such as hares, partridges, and pheasants, were reserved for the use and enjoyment of the rich. The Game Act of 1671 directed that game could be taken only by people possessing freehold property worth at least £100 per year, by those holding leaseholds of 99 years or longer or copyholds worth at least £150 per year, or by those who were the sons or heirs apparent of esquires or others "of higher degree." Not only did this act restrict the right to hunt to less than one percent of the population, it also restricted it specifically to the established landed rich. In 1707 the penalty for poaching was increased to a blanket fine of £5 or three months in prison. In 1723 the Black Act, "for the more effectual punishment of wicked and evildisposed persons going armed, in disguise" ensured that merely appearing in the vicinity of a game reserve, armed and with face blackened, was a hanging offence. In 1773 a new Night Poaching Act raised the fine up to £50, depending on the number of prior convictions, and allowed imprisonment of up to 12 months plus public whipping. Williamson, *Polite Landscapes*, 136.

5. In the eighteenth century, the Chesapeake woods were alive with grouse, woodcock, squirrels, rabbits, wild turkey, and deer. Panthers, cougars, bobcats, wolves, and bears lived in the hills to the west of the bay. On May 17, 1796, hunters rendezvoused at Irvine's Lick in Kentucky and killed 7,941 squirrels in a single day. Thomas D. Clark, *Frontier America* (New York: Charles Scribners Sons, 1969), 221. For more information about Maryland hunting

see Meshach Browning's *Forty-Four Years of the Life of a Hunter* (Philadelphia: J. B. Lippincott, 1860).

6. Thomas Jefferson, *Thomas Jefferson's Garden Book*, ed. Edwin Morris Betts (Philadelphia: American Philosophical Society, 1944), 322–23.

7. Edward C. Carter II, John C. Van Horne, and C. E. Brownell, eds., *Latrobe's View of America, 1795–1820: Selections from the Watercolors and Sketches* (New Haven: Yale University Press, 1985), 72.

Chapter 1   A Craftsman's Garden

1. For the past two decades, landscape architect Arthur A. Shurcliff has been criticized for creating elaborate town gardens at the homes of merchants and craftsmen for the restoration of Colonial Williamsburg. This diary of an eighteenth-century artisan may help to counteract some of this criticism, which includes Charles B. Hosmer Jr., "The Colonial Revival in the Public Eye: Williamsburg and Early Garden Restoration," in Alan Axelrod, ed., *Colonial Revival in America* (New York: Norton, for the Henry Francis du Pont Winterthur Museum, 1985), 53–63.

For additional information on the life and family of William Faris, see Lockwood Barr, "William Faris, 1728–1804" *Maryland Historical Magazine*, 36 (1941): 420–39; Barr, "Family of William Faris (1728–1804)" *Maryland Historical Magazine* 37 (1942): 423–32; William Faris's Diary, 1792–1804, MS 2160, Maryland Historical Society (hereafter MHS), Baltimore; William Faris–William McParlin Account Books, MS 353, MHS; Faris Papers, MS 343, MHS; William Faris Account Books, MS 1104, MHS; Faris Design Book, MS 348, MHS.

2. Rev. Andrew Burnaby, *Travels Through the Middle Settlements in North-America in the Years 1759–1760* (London, 1775), 65; "Journal of Lord Adam Gordon, Journal of an Officer who travelled in America and the West Indies in 1764 and 1765," in *Travels in American Colonies*, ed. Newton D. Mereness (New York, 1916), 408; letter from Lord Baltimore to Rev. Bennet Allen, Jan. 1767, Calvert Papers, MS 174, MHS; John Ferdinand D. Smyth, *A Tour In the United States of America* (London, 1784), vol. 2, p. 185.

For discussions of the economic rise and fall of Annapolis see Edward C. Papenfuse, *In Pursuit of Profit: The Annapolis Merchants in the Era of the American Revolution, 1763–1805* (Baltimore: Johns Hopkins University Press, 1975); Ronald Hoffman, *The Spirit of Dissention: Economics, Politics, and the Revolution in Maryland* (Baltimore: Johns Hopkins University Press, 1976); Jean B. Russo, "The Structure of the Anne Arundel County Economy," in *Annapolis and Anne Arundel County, Maryland: A Study of Urban Development in a Tobacco Economy, 1649–1776*, ed. Lorena Walsh (MS on file at Historic Annapolis, Inc., Annapolis, Md.). This period is also discussed in two earlier histories: David Ridgely, *Annals of Annapolis* (Baltimore: Cushing & Brother, 1841); and Elihu Riley, *The Ancient City: A History of Annapolis in Maryland, 1649–1887* (Annapolis: Record Printing Office, 1887).

3. Bernard M'Mahon, *The American Gardener's Calendar* (Philadelphia: B. Graves, 1806), 76.

4. Information on Faris's property from Jane McWilliams and Edward C. Papenfuse, *Appendix F, Lot Histories and Maps*, vol. 2, NEH Grant #H 69-0-178, 000014–000032, Annapolis, 1971, at the Maryland State Archives (hereafter MSA), Annapolis. Additional history of Faris's land purchases and property improvements may be found in the following: Historic Annapolis Index Files, NEH Grants RS 0067-79-0738 and RS 20199-81-0738), MSA; Annapolis 1798 Federal Direct Tax List, MSA; Annapolis Records, Mayor's Court Min-

utes 1766–1772, #5099:180, Oct. 28, 1766; *Maryland Gazette*, Oct. 5, 1769. Information about Faris's wall and gate from Faris's Diary: Apr. 3, 1794; Mar. 31, 1792; Apr. 7, 1804. Faris-McParlin Account Books (MS 353), June 17, 1799; Nov. 16, 1799. For a further contemporary description of Faris, see a poem, "The Will of William Faris," written in 1790 by Charlotte Hesselius, a friend of Faris's daughters. It appeared in *Scribner's Monthly* in January 1879 in "Old Maryland Manners," by Francis B. Mayer; and it was reprinted by Riley in *The Ancient City* in 1887. Charlotte Hesselius (1770–1794) was the daughter of the portrait painter John Hesselius (1728–1778) and his wife Mary (1739–1820). In his diary, Faris reported Charlotte's marriage to Thomas Jennings Johnson Jr. (1766–1807), in Annapolis in 1792. She died at age 24, only five years after she wrote the poem. (J. Hall Pleasants Files, MHS).

5. Charles Carroll of Carrollton to Charles Carroll of Annapolis, Oct. 30, 1769; Sept. 29, 1774, Carroll Papers, MS 206, MHS; *Virginia Gazette*, Jan. 6, 1767.

6. Faris's Diary: Mar. 25, 1792; May 1, 1803.

7. Faris's tulip plantings by year, as reported in his diary on Nov. 5, 1798:

1792:	1490	1799:	2425
1795:	1900	1801:	3945
1797:	1956	1803:	3566
1798:	1645	1804:	2339

8. The neighbor with the narcissus bulbs was Henri Joseph Stier (1743–1821), a Belgian nobleman who had fled Antwerp across the Dutch border to Amsterdam in 1794. The Philadelphia *General Advertiser* of Oct. 13, 1794, announced that the Stier party had just arrived on the *Adriana*. In the fall of 1795, Stier came to Maryland and rented the estate Strawberry Hill, near Annapolis, which had garden terraces dropping toward the Severn River. In the fall of 1797, the family moved again, this time to the Jennings-Paca House. John V. L. McMahon, *An Historical View of the Government of Maryland* (Baltimore: F. Lucas Jr., Cushing & Sons and William and Joseph Newl, 1831), 434, n. 39.

9. Carroll Papers, MS 206, Charles Carroll of Carrollton to Edmund Jennings, Oct. 14, 1766.

10. Faris Diary, May 8, 1792. According to his diary, Faris sold tulip bulbs to Alexander Contee Hanson (Apr. 18, May 4, and Dec. 4, 1801; Apr. 25, 1796; Apr. 27, May 3, 1797; Apr. 21, May 16, 1799; Apr. 17, 1800); Thomas Harwood (May 8, June 13–14, 1792); and John O'Donnell (May 6, June 13–14, 1792). He traded plants for plants with Dr. Upton Scott (May 5, 1792); Charles Wallace (Mar. 23, 1792); and Maximillian Heuisler (May 14, 1793).

Marking plants with notched sticks was the custom of the period in the Chesapeake. Thomas Jefferson notching the plant itself or a stick: "Near the tip end of every plant cut a number of notches which will serve as labels, giving the same number to all plants of the same species. Where the plant is too small to be notched, notch a separate stick & tye it to the plant. Make a list on paper of the plants by their names & number of notches." Jefferson, *Garden Book*, 117.

11. By 1769, the work on his external surroundings was largely completed, and Faris advertised to rent out his gardener to neighbors, who were also beginning to order their outdoor environments for style as well as utility. *Maryland Gazette*, Oct. 5, 1769.

12. M'Mahon, *American Gardener's Calendar*, 73–76.

13. Ibid. Some American garden historians claim that boxwood was not commonly used as bed edging until well after the Revolution, but in the *South Carolina Gazette*, Mar. 14, 1768,

Martha Logan advertised for sale "A Fresh assortment of very good garden seeds and flower roots . . . and box for edging beds."

14. Faris's Diary, Mar. 10, 1799; Hesselius, "The Will of William Faris."

15. In 1665 John Rea warned gardeners, "the roots if not cut away on the inside with a keen spade every other year, will run into the beds, and draw from the flowers much of their nourishment." John Rea, *Flora, Ceres, Pomona* (London: J. G. for Richard Marriott, 1665).

16. Barr, "William Faris, 1728–1804," 425.

17. [John Beale Bordley], *Gleanings from the Most Celebrated Books on Husbandry, Gardening, and Rural Affairs* (Philadelphia: James Humphreys, 1803), 184.

18. On July 26, 1799, Faris reported in his accounts, "Cash paid to Syrus (Johnson) for Building the porch and Steps £1/10/1." Faris-McParlin Account Books (MS 353).

19. Carroll Papers, MS 206, Charles Carroll of Annapolis to CC of Carrollton, May 4, 1770; 1798 Particular Tax List, Baltimore and Annapolis, MSA; *Pennsylvania Gazette,* Jan. 30, Mar. 27, 1782.

20. *Maryland Gazette,* Sept. 20, 1759; Feb. 2 and 23, 1764.

21. In 1577, the English author Thomas Hill wrote, "The gardener possessing a Pump in his ground, or fast by, may with long and narrow troughes well direct the water unto all beds of the Garden, by the pathes between, in watering sufficiently the roots of all such herbs, which require much moisture." Hill, *The Gardener's Labyrinth* (1652), ed. Richard Mabey (London: Oxford University, 1987), 83. Faris's Diary: Apr. 8, 1793; Mar. 26, 1794; June 17, 1796; Apr. 3, 1800; Apr. 7, 1800.

22. Faris's Diary, Mar. 21, 1793; *Baltimore Daily Repository,* Jan. 9, 1791.

23. Faris's Diary, Mar. 21, 1798; [Bordley], *Gleanings,* 98.

24. Richard Parkinson, *A Tour in America in 1798, 1799, and 1800,* (London: J. Harding, 1805), 79.

25. Faris's Diary: Mar. 23, 1793; Feb. 19, 1794; May 9, 1794; May 31, 1796; and Parkinson, *A Tour in America,* 185.

26. Faris's Diary, Mar. 9 and 15, 1792.

27. Parkinson, *A Tour in America,* 185.

28. See Charles Peale Polk's painting of Margaret Baker Briscoe, MHS; George Wiedenbach's of Belvedere, MHS; and chairback paintings by Francis Guy of Mount Clare and Willow Brook at the Baltimore Museum of Art.

29. John Beale Bordley, *Essays and Notes on Husbandry and Rural Affairs* (Philadelphia: Thomas Dobson, 1799), 391.

30. Johann David Schoef, *Travels in the Confederation, 1783–1784,* ed. and trans. Alfred J. Morrison (New York: Burt Franklin, 1968) vol. 1, pp. 93, 60.

31. Annapolis 1798 Federal Direct Tax List, MSA.

32. [Bordley], *Gleanings,* 187.

33. Carroll Papers, MS 206, Charles Carroll of Annapolis to CC of Carrollton, Nov. 26, 1773. *Maryland Gazette,* Aug. 21, 1760; Apr. 28, 1763; and Mar. 15, 1764.

34. Thomas Hughes, *A Journal by Thos. Hughes,* introduced by E. A. Benians (Cambridge: Cambridge University Press, 1947), 19; William Hugh Grove, "Virginia in 1732: The Travel Jour-

nal of William Hugh Grove," ed. Gregory A. Stiverson and Patrick H. Butler III, *Virginia Magazine of History and Biography* 85 (1977): 35. Francis Louis Michel, "Report of the Journey of Francis Louis Michel from Berne, Switzerland, to Virginia, October 2, 1701–December 1, 1702," ed. and trans. William J. Hinke, *Virginia Magazine of History and Biography* 24 (1916): 33.

35. Captain John O'Donnell (1749–1805) was an Irish sea captain whose vessel, "Pallas," arrived in Baltimore in March 1785 and was the first to bring goods to the city directly from China. He amassed a fortune trading with China and India. Englishman Richard Parkinson found the grounds surrounding O'Donnell's country seat "a very handsome garden in great order, a most beautiful greenhouse and hot house . . . a very magnificent place for that country." Thomas Twinning, a British official who visited O'Donnell in Baltimore in 1795, noted that O'Donnell's travels in India were reflected in the design: "The long, low house with a deep veranda in front had somewhat the appearance of a pucka bungalow." O'Donnell bought plants from Faris. Parkinson, *A Tour in America*, 77; Thomas Twinning, *Travels In America 100 Years Ago* (New York: Harper & Brothers, 1894), 118, 125, 126.

36. Parkinson, *A Tour in America*, 221.

37. Dr. Upton Scott came to Annapolis in 1753 as physician to Colonial Governor Horatio Sharpe. Scott built his house and began its gardens about 1765. Dr. Scott's hobby was botany, and he built large formal gardens extending to Duke of Gloucester Street in the front and to Spa Creek in the rear. He also maintained a 14-by-30-foot greenhouse filled with rare plants and shrubs. Faris and Scott exchanged hundreds of flower and vegetable seeds and plants. Upton Scott lived in a two-story brick home (54' × 45'). His grounds contained the greenhouse, a brick carriage house (14' × 14'), a brick smokehouse (10' × 10'), a brick stable and cow house (30' × 20'), and a brick poultry house (10' × 6 '). Annapolis 1798 Federal Direct Tax List, MSA; Charles Willson Peale, *The Selected Papers of Charles Willson Peale*. "Charles Willson Peale: Artist in Revolutionary America, 1735–1791," ed. Lillian B. Miller, published for the National Portrait Gallery, Smithsonian Institution (New Haven: Yale University Press, 1983), 1:148; Papenfuse, *In Pursuit of Profit*, 18, 78, 144, 154; *St. Anne's Parish Register and Vestry Proceedings*, 2:21; Edith Rossiter Bevan, "Gardens and Gardening in Early Maryland," *Maryland Historical Magazine* 50 (1950), 243–70; Bevan, "Journal of a Voyage," *Maryland Historical Magazine* 2 (1916), 132; Historic Annapolis Index Files [NEH Grants, RS 0067-79-0738 and RS 20199-81-0738], MSA). Faris's Diary, Nov. 15, 1801. David Baillie Warden, "Journal of David Baillie Warden, Secretary to the American Legation in Paris" *Maryland Historical Magazine* 11:129.

38. Faris's Diary, Feb. 21, 1804.

39. Ibid.

40. Faris's Diary, Apr. 6, 1795. Martins may not have been necessary for the success or failure of Faris's gardening efforts, but they certainly made the task of working outdoors more comfortable. Near the Chesapeake Bay, mosquitoes seriously diminished the pleasures of colonial gardening; and their predators, the martins, were welcome guests. In order to encourage them to remain close to a garden, pottery nesting bottles were sometimes hung beneath the eaves of buildings. The lead-glazed pottery bird bottles of the eighteenth century were suspended from a nail whose head fitted into the cutout base. At the time of his death in 1746, John Burdett of Williamsburg, Virginia, owned sixteen "bird bottles" and may have used them in groups to attract the gregarious birds. "Martin pots" were offered for sale in 1752 at a store near the Bruton Church in Williamsburg. The 1767 inventory of Philip Ludwell of Green Spring, Virginia, included "Martin Potts." See A. N. Hume, *Archae-*

ology and the Colonial Gardener, Colonial Williamsburg Archaeological Series No. 7 (Williamsburg: Colonial Williamsburg Foundation, 1974), 67.

Chapter 2   Gardens of the Gentry

1. John Adams, *The Works of John Adams*, ed. C. F. Adams (Boston, 1865), diary entry of Feb. 17, 1777; cartographer Charles Varle and engraver Francis Shallus, Warner and Hanna's "Plan of the City and Environs of Baltimore, Respectfully dedicated to the Mayor, City Council & Citizens thereof by the Proprietors," 2nd ed. (Baltimore, 1801; 1st 1799, drawn 1797) (engraving, 50 × 72.5 cm.) (hereafter Warner and Hanna Plan of Baltimore).

   For a history of the ownership of Chatsworth, see Chapter 9, note 23.

2. Illustration is "Samuel Harrison's Landing Near Herring Bay," mantle painting, oil ca. 1730, private collection, Maryland Historical Trust file AA-268; Historic American Building Survey, MD-284.

3. Jacobean-style Bacon's Castle, built before 1676, is Virginia's oldest known brick dwelling. The central walk in the garden at Bacon's Castle sits on a north-south axis and is 12 feet wide. Six large kitchen garden beds, each 74 feet by about 98 feet, are crossed by 8-foot-wide walkways and surrounded by an outer walk that is 10 feet wide. Garden archaeology by Nicholas M. Luccketti.

4. See James L. Reveal, *Gentle Conquest: The Botanical Discovery of North America with Illustrations from the Library of Congress* (Washington, D.C.: Starwood Publishing, 1992). William Byrd, *The London Diary (1717–1721) and Other Writings*, ed. Louis B. Wright and Marion Tinling (New York: Oxford University Press, 1958), 393.

5. Carroll Papers, MS 206, Charles Carroll of Carrollton to CC of Annapolis, Oct. 26, 1774.

6. Charles Willson Peale painting on loan to Maryland Historical Society from Johns Hopkins University Peabody Collection.

7. Annapolis 1798 Particular Tax List, MSA; Kenneth G. Orr and Ronald G. Orr, "The Archaeological Situation at the William Paca Garden, Annapolis, Maryland: The Spring House and the Presumed Pavilion House Site" (typescript on file, William Paca Garden Visitors' Center, Annapolis, Md., 1975); Bruce B. Powell, "Archaeological Investigation of the Paca House Garden, Annapolis, Maryland. November 16, 1966" (typescript on file, William Paca Garden, Annapolis, Md.); Mark Leone, "Interpreting Ideology in Historical Anthropology: Using the Rules of Perspective in the William Paca Garden in Annapolis, Maryland," in *Ideology, Power, and Prehistory*, ed. Daniel Miller and Christopher Tilley (New York: Cambridge University Press, 1984), 25–35.

8. Hannah Callender, "Extracts from the Diary of Hannah Callender," *Pennsylvania Magazine of History and Biography* 12 (1888): 454–55.

9. M'Mahon, *American Gardener's Calendar*, 76.

10. Carroll Papers, MS 203, Charles Carroll of Carrollton to William Graves, Sept. 15, 1765.

11. Charles Willson Peale Diary, June 11, 1804, Manuscripts Collection, American Philosophical Library, Philadelphia; Baltimore Museum of Art, *"Anywhere So Long As there be Freedom," Charles Carroll of Carrollton, His Family & His Maryland*, catalogue of an exhibition (Baltimore: Baltimore Museum of Art, 1975); Carroll Papers, MS 206: Charles Carroll of Carrollton to Shields & Carvier, Seedsmen, Sept. 20, 1773; CC of Carrollton to William Graves, Sept. 15, 1765. For detailed description of the Carroll garden on Spa Creek, see Elizabeth

Kryder-Reid, *Landscape as Myth: The Contextual Archaeology of an Annapolis Landscape* (Ph.D diss., Brown University, 1991).

12. *South Carolina Gazette,* May 22, 1749.

13. Anne Grant, *Memoirs of an American Lady, with Sketches of Manners and Scenery in America, as They Existed Previous to the Revolution* (New York, 1809), 26, 21–22.

14. Karen Madsen, "William Hamilton's Woodlands," unpublished manuscript, 1988, 7.

15. Elizabeth Drinker, *Diaries 1759–1807,* Nov. 4, 1803, Manuscript Collection, Historical Society of Pennsylvania, Philadelphia.

16. Walter Kendall Watkins, "The Hancock House and Its Builder," *Old Time New England* 17 (July 1926): 7.

17. Louis R. Caywood, "Excavations at Green Spring Plantation," (Yorktown, Va., 1955); Jane Carson, "Green Spring Plantation in the Seventeenth Century" (December 1954), Special Collections, Colonial Williamsburg Foundation Library.

18. William Byrd II, *The Writings of Colonel William Byrd,* ed. John Spenser Bassett (New York, 1970), 357–58; *Virginia Gazette,* Feb. 15, 1770.

19. William M. Kelso, "Landscape Archaeology: A Key to Virginia's Cultivated Past," in *British and American Gardens in the Eighteenth Century,* ed. Robert P. Maccubbin and Peter Martin (Williamsburg, 1984), 162–63; Camille Wells, "Kingsmill Plantation: A Cultural Analysis" (M.A. thesis, University of Virginia, 1976); Mary R. M. Goodwin, "Kingsmill Plantation" (Sept. 1958) Special Collections Colonial Williamsburg Foundation Library.

20. Kryder-Reid, *Landscape as Myth,* 204; Jack P. Greene, ed., *The Diary of Colonel Landon Carter of Sabine Hall, 1752–1778,* 2 vols., (Charlottesville: University Press of Virginia, 1965) 1:254.

21. *Virginia Herald,* Jan. 24, 1800; ibid., Oct. 26, 1803; *Virginia Argus,* July 24, 1805. Other eighteenth-century terraced gardens in Virginia probably included Belle Isle in Lancaster County, Belmont in Stafford Co., Berkley in Charles City Co., Castle Hill in Albemarle Co., Cherry Grove in Nansemond Co., The Chimneys in Frederick Co., Claremont in Surry Co., Cleve in King George Co., Delk Farm in Smithfield in Isle of Wight Co., Elmwood and Font Hill in Essex Co., Elsing Green in King William Co., Eppington in Chesterfield Co., Fairfield in Clarke Co., Federal Hill in Frederick Co., Four Mile Tree in Surry Co., Gaymont in Caroline Co., Grovemont in Richmond Co., Hickory Hill in Ashland in Hanover Co., Mount Airy in Warsaw in Richmond Co., Mount Pleasant and Pleasant Point in Surry Co., Olive Hill on the Appomattox River, Salubria in Culpepper Co., Shoal Bay in Smithfield, Strawberry Plain in Isle of Wight Co., Wigwam in Nansemond Co., Wilton in Middlesex Co., and Woodbury Forest in Madison Co.

22. *Virginia Gazette,* May 2, 1777, and Feb. 5, 1780.

23. Nineteenth-century gardens were built at Randolph Harrison's Elk Hill, John Tabb's White Marsh, Hampstead in New Kent Co., James Madison's Montpelier, David Higgenbothan's Morven in Albemarle Co., Richard Bowen's Mirador in Albemarle Co., James C. Bruce's Berry Hill in Halifax Co., and John and Robert Wilson's Dan's Hill in Pittsylvania Co.

24. John Harris, *The Artist and the Country House,* (London: Sotheby Parke Bernet Publications, 1985), plates 189a, 236, 246; Laurence Fleming and Alan Gore, *The English Garden* (London, 1980), 67. William Lawson, *A New Orchard and Garden* (London, 1618).

25. Benjamin Henry Latrobe, *The Virginia Journals of Benjamin Henry Latrobe, 1795–1798*, ed. Edward C. Carter II (New Haven: Yale University Press, 1977), 473.

26. Burnaby, Andrew. *Travels Through the Middle Settlements in North-America in the Years 1759 and 1760* (1775; Ithaca, N.Y.: Cornell University Press, 1976), 65–67. Lord Adam Gordon. "Journal of an Officer who Travelled in America and the West Indies in 1764 and 1765," in *Travels in the American Colonies*, ed. Newton Mereness (New York: Macmillan, 1916), 408.

27. William Eddis, *Letters from America*, ed. Aubrey C. Land (Cambridge: Harvard University Press, 1969), 12.

28. Some believe that because Captain Charles Ridgely began to build his mansion in 1783 and died in 1790, the year it was completed, most of the credit for Hampton's gardens and grounds should be given to his nephew Charles Carnan Ridgely, who inherited the property and lived there nearly forty years, until he died in 1829.

29. Alexander Samuel Salley, ed., *Narratives of Early Carolina 1650–1708* (New York: Charles Scribner's Sons, 1911), 100. Byrd, *The London Diary*, 526.

30. Lynn Dakin Hastings, *Hampton National Historical Site* (Towson, Maryland: Historic Hampton, Inc., and National Park Service, U.S. Department of the Interior, 1986) 58, 64, 69.

31. Warner and Hanna Plan of Baltimore.

32. Francis Guy painting of Beech Hill (ca. 1805), Baltimore Museum of Art (hereafter BMA); 1798 Particular Tax List, Baltimore County; *Baltimore News,* Jan. 28, 1848; *Baltimore Sun,* July 25, 1914 and Mar. 1, 1936; Alice B. Lockwood, *Gardens of Colony and State* (New York: Charles Scribner and Sons, 1931), 2:12; Robert Gilmore II, *Reminiscences* (Baltimore, 1844) MS, MHS.

33. Francis Guy paintings of Perry Hall at the MHS; *Maryland Journal & Baltimore Advertiser,* Apr. 6, 1774; Edith Rossiter Bevan, "Perry Hall: Country Seat of the Gough and Carroll Families," *Maryland Historical Magazine* 45 (1950): 33–46.

34. *Baltimore Daily Intelligencer,* Apr. 22, 1794.

35. *Baltimore News,* Nov. 27, 1822; *American and Commercial Advertiser,* Mar. 2, 1824; *Baltimore Sun,* Mar. 1, 1936.

36. When cutting their slopes and flats, colonial gardeners were careful to preserve the top soil for their later plantings. When Charles Carroll of Carrollton was leveling and building terraces at his Annapolis property, his father advised, "In levelling your ground I hope you have been Carefull to preserve the Top Soil & to lay it on again, Sowe your Clover seed when the soil is moist. Rake it & when pretty dry Role it with your Garden Roler if not Too Heavy." Carroll Papers, MS 206, Charles Carroll of Annapolis to CC of Carrollton, Aug. 23, 1771.

37. "The Chevalier D'Annemours," *Maryland Historical Magazine* 1 (1906): 241–46.

38. Claypole's *Daily Advertiser* (Philadelphia), Aug. 17, 1792.

39. Abigail Adams, *Letters of Mrs. Adams, the Wife of John Adams*, ed. Charles Francis Adams (Boston, 1915), 420.

40. François, duc de La Rochefoucauld-Liancourt, *Travels through the United States of North America in the Years 1795, 1796 and 1797*, 4 vols. (London, 1799) 2:438.

41. Warner and Hanna Plan of Baltimore.

42. 1798 Particular Tax List, Baltimore County. Ella K. Barnard, "Mount Royal and Its Owners," *Maryland Historical Magazine* 26 (1931): 311–17.

43. Warner and Hanna Plan of Baltimore; *Federal Gazette,* Nov. 6, 1821; *Baltimore Sun,* July 25, 1914; *Baltimore Evening Sun,* Apr. 9, 1940.

44. See Francis Guy paintings of Bolton, MHS and BMA.

45. *Journal of Jasper Danckaerts, 1679–1680,* in *Original Narratives of Early American History,* J. Franklin Jameson, gen. ed. (New York: Charles Scribner's Sons, 1913), 107.

46. Manasseh Cutler, *Life, Journal, and Correspondence of Rev. Manasseh Cutler, LL.D.,* ed. William Parker and Julia Perkins Cutler (Cincinnati, 1888), 275.

47. Juliana Margaret Conner, Diary 1827, Southern Historical Collection, University of North Carolina Library, Chapel Hill.

48. Warner and Hanna Plan of Baltimore.

49. In the late 1790s a merchant, Jeremiah Yellot, bought Ghequiere's home and named it Woodville. The estate changed hands again almost immediately, acquired by an Irish merchant, Hugh McCurdy, who renamed the elegant property Grace Hill. See Francis Guy paintings of Woodville and Grace Hill, BMA; Warner and Hanna Plan of Baltimore; *Federal Gazette & Baltimore Daily Advertiser,* Apr. 5, 1800.

50. My information on the Howard statues comes from a personal conversation with Stiles Tuttle Colwill, former curator of the Maryland Historical Society, whose knowledge of Baltimore country seats is encyclopedic.

51. Lockwood, *Gardens of Colony and State,* 27.

52. Ezra Stiles, "Ezra Stiles in Philadelphia, 1754," *Pennsylvania Magazine of History and Biography* 16 (1892): 375; Daniel Fisher, "Extracts from the Diary of Daniel Fisher, 1755," *Pennsylvania Magazine of History and Biography* 17 (1893): 267.

53. Callender, "Extracts from the Diary," 454–55.

54. Robert Bolling, "An Incitation to Vineplanting" (poem), 1772, Brock Collection, BR 64, Huntington Library, San Marino, California.

55. Philip Vickers Fithian, *Journal & Letters of Philip Vickers Fithian, 1773–1774: A Plantation Tutor of the Old Dominion,* ed. and introd. Hunter D. Farish. (Williamsburg, Va.: Colonial Williamsburg, Inc., 1943), 94–95.

56. Lockwood, *Gardens of Colony and State.* 2:34.

57. *Pennsylvania Packet,* Sept. 12, 1796.

58. Painting of Mrs. Gerrard Briscoe by Charles Peale Polk, MHS.

59. *New York Daily Advertiser,* July 2, 1800; ibid., June 24, 1805.

60. Francis Guy chairback painting of Belvedere, BMA; Weidenback painting of Belvedere, MHS; Warner and Hanna Plan of Baltimore.

61. Moreau de St. Mery, *American Journey,* 88; La Rochefoucault-Liancourt, *Travels Through the United States,* 258, 580.

62. Rosemary Verey, *Classic Garden Design* (New York: Congdon & Weed, 1984), 11; Elizabeth Pryor, *Flowers and Flowering Bushes in the Colonial Chesapeake,* National Colonial Farm Research Report, No. 17, (Accokeek, Md.: Accokeek Foundation, 1984), 6.

63. Twinning, *Travels In America,* 288. Moreau de St. Mery, *American Journey,* 79.

64. Charles Willson Peale's miniature paintings of Dr. Henry Stevenson and Parnassas, MHS; Warner and Hanna Plan of Baltimore; *Maryland Gazette,* Mar. 25, 1784.

65. Warner and Hanna Plan of Baltimore; "Rose Hill," *Maryland Historical Magazine* 6 (1911): 23.

66. Edith Rossiter Bevan, "Willow Brook, Country Seat of John Donnell," *Maryland Historical Magazine* 44 (1949): 33–41; Francis Guy painting of Willow Brook, BMA.

67. *Federal Gazette,* Apr. 18, 1800. Visiting hot springs and having bathhouses for cold baths at home for medicinal purposes became popular in the Chesapeake during the eighteenth century. When Charles Carroll of Carrollton planned his cold bath in 1778, he wrote to his father to have stone masons prepare enough stone to line a bath 10 feet long, 8 feet wide, and 4½ feet deep. Carroll Papers, MS 206, Charles Carroll of Carrollton to CC of Annapolis, May 24–26, 1778. In 1728 William Byrd II visited Harvey Harrison's in Virginia, where he reported a cold spring bath 5 feet deep and a 5-foot-square bathhouse in which Harrison, "formerly troubled both with the Gripes & the Gout, fancies he receives benefit by plunging every day in cold Water." Wendy Martin, ed., *Colonial American Travel Narratives* (New York, Penguin Books, 1994), 122.

68. Francis Guy paintings of Mount Clare, BMA; copy of a 1770s painting of Mount Clare by an unknown artist, at Mount Clare Mansion, Baltimore.

69. Mary Ambler, "Diary of M. Ambler, 1770," *Virginia Magazine of History and Biography* 45 (1937): 168–70.

70. Otho Williams to George Washington, Oct. 7, 1789, Otho Williams Papers, MS 908, MHS.

71. John Adams, *Works,* diary entry for Feb. 23, 1777.

Chapter 3   The Republican Garden

1. Jedidiah Morse, *The American Geography; or, a View of the Present Situation of the United States of America* (Elizabethtown, N.J.: Sheppard Kollock, 1789), 296.

2. Peale Diary, June 11, 1804.

3. List of plants Charles Carroll ordered for his garden from Messrs Wallace Johnson & Muir Merchants, from Sally Mason, Carroll Papers project, supervised by Ronald Hoffman, the Omohundro Institute for Early American History and Culture, Williamsburg, Va.

4. For a detailed discussion see Tom Williamson's *Polite Landscapes.*

5. Warner and Hanna Plan of Baltimore.

6. Ibid.

7. Carroll Papers, MS 206, Charles Carroll of Annapolis to CC of Carrollton, Sept. 17, 1772; Apr. 10, 1775.

8. Dorsey's House and Garden Plans, Architectural Drawings Collection, MHS.

9. Carroll Papers, MS 206, Charles Carroll of Annapolis to CC of Carrollton, Mar. 27, 1777; Henry Skipwith to John Hartwell Cocke, Mar. 19, 1813, Cocke Family Papers, Alderman Library, University of Virginia, Charlottesville; Thomas Jefferson to Pierre Charles L'Enfant, Apr. 10, 1791, in *The Papers of Thomas Jefferson,* ed. J. P. Boyd, 27 vols., (Princeton: Princeton University Press, 1982), 20:462.

10. Dorsey's House and Garden Plans.

11. Byrd, *London Diary*, 428.

12. Ibid; Charles Carroll of Carrollton, letter of Apr. 9, 1823, Sotheby's *Fine Printed and Manuscript Americana*, Sale #5700, Item #56, Apr. 16, 1988.

13. Dorsey's House and Garden Plans.

14. Ibid.

15. Bordley, *Essays and Notes*, 4.

16. M'Mahon, *American Gardener's Calendar*, 73.

17. Fithian, *Journal*, 79–82.

18. Warner and Hanna Plan of Baltimore: country seats of Gilbert Bigger, Dr. James McHenry, Christian Raborg, Henry Stouffer, Major Thomas Yates. Also see Francis Guy's paintings of the homes of General William Buchanan, John Donnell, George Grundy, Levi Pierce, Hugh McCurdy, and Jeremiah Yellot at BMA and MHS.

19. Murraywhaithe Collection, Scottish Record Office, Edinburgh, GO219/284/5.

20. Callender, Diary. 454–55.

21. George Washington, *The Diaries of George Washington*, ed. John C. Fitzpatrick (New York: Houghton Mifflin, 1925), 341, 343.

22. Lewis Beebe, Journal (1776–1801), MS Collection, Historical Society of Pennsylvania, Philadelphia, spring 1800. A rod equals $16\frac{1}{2}$ feet.

23. Deborah Norris Logan, *The Norris House.*(Philadelphia, 1867), 5–6. Beebe, Journal, spring 1800.

24. Joseph Prentis Garden Book, spring 1786, MS Collection of Colonial Williamsburg Foundation Library; Faris's Diary, Mar. 19, 1792.

25. 1798 Particular Tax List, Baltimore County.

26. Fithian, *Journal*, 79–82.

27. Bordley, *Essays and Notes*, 76, 77, 79.

28. Philadelphia *Mercury*, Apr. 10, 1740.

29. The deer population had diminished by mid-century. Colonial Secretary William Eddis wrote from Annapolis on June 8, 1770 (*Letters from America*, 32):

> Deer a few years since were very numerous in the interior settlements; but from the unfair methods adopted by the hunters their numbers are exceedingly diminished. These people, whose only motive was to procure the hide of the animal, were dexterous, during the winter season, in tracing their path through the snow; and from the animal's incapacity to exert speed under such circumstances, great multitudes of them annually slaughtered and their carcasses left in the woods. This practice, however, has been thought worthy of the attention of the legislature, and an act of assembly has taken place, laying severe penalties on persons detected in pursuing or destroying deer.

30. William Gooch Transcripts, Virginia Historical Society, Richmond; Daniel Fisher, "Narrative of Daniel Fisher" in *Some Prominent Virginia Families*, ed. Louise Pecquet du Bellet

(Lynchburg, Va.: J. P. Bell, 1907); Morse, *The American Geography*, 381; Isaac Weld, *Travels Through the States of North America and the Provinces of Upper and Lower Canada During the Years 1795, 1796, 1797* (London, 1799), 53; Kate Mason Rowland, *The Life of George Mason, 1725–1792, Including His Speeches, Public Papers and Correspondence* (New York: Russell & Russell, 1964), 98.

31. Bordley, *Essays and Notes*, 79.

32. Rosalie Stier Calvert, *Mistress of Riversdale: The Plantation Letters of Rosalie Stier Calvert, 1795–1821*, ed. Margaret Law Callcott (Baltimore: Johns Hopkins University Press, 1991), 196.

33. Ibid., 181.

34. Peter Kalm, *Peter Kalm's Travels in North America: The English Version of 1770*. ed. Adolph B. Benson (New York: Dover, 1987), 161–62; Bordley, *Essays and Notes*, 258. Bordley's comments on stocking a fishpond included the following (*Gleanings*, 124, 259):

> 1. Carp . . . will not thrive in a cold water
> 2. Tench . . . pond should have a muddy bottom with weeds
> 3. Perch . . . like a clear and moderately deep water
> 4. Crucian . . . brought from Germany
> 5. Gold and Silver Fish . . . possessing a finer flavor
> 6. Pike . . . pond . . . of good depth, with weeds growing in it
> 7. Eels . . . never breed in perfect standing water
> 8. Bream . . . Roach . . . Dace . . . Minnows . . . kept in ponds with Pike and Perch, as food for them . . . Ruff or Pope, which is much like the Perch, but esteemed better eating: and the Gudgeon . . . equal in goodness to the delicate Smelt . . . delights in a gravelly bottom.

35. *Federal Gazette*, June 14, 1800; Warner and Hanna Plan of Baltimore; Nicolino V. Calyo, 1834 painting of Harlem, in the collection at Winterthur; Dieter Gunz, *The Maryland Germans* (Princeton: Princeton University Press, 1948), 155–81; *Baltimore News*, May 5, 1872; *Baltimore News Post*, June 13, 1941; *Baltimore Sun*, July 25, 1914; *Horticulturalist* 7 (1857): 350–52.

36. Carroll Papers, MS 206, Charles Carroll of Carrollton to CC of Annapolis, Oct. 26, 1774; Baltimore County 1798 Particular Tax List, MSA; Elizabeth B. Anderson, *Annapolis: A Walk through History* (Centreville, Md.: Tidewater Publishers, 1984), 58–59, 122–23; *Federal Gazette*, June 14, 1800; Sept. 22, 1750; A traveler, James Birket, wrote of Capt. Godfrey Malbone's, at Newport, R.I., "Handsome Garden with variety of wall fruits And flowers." James Birket, *Some Cursory Remarks (Made by James Birket in his Voyage to North America 1750–1751)* (New Haven: Yale University Press, 1916), 27.

37. Thomas Anburey, *Travels through the Interior Parts of America*, 2 vols. (New York: New York Times, 1969), 2:323–24.

38. Baltimore *Federal Gazette*, Aug. 11, 1800.

39. Eddis, *Letters from America*, 14.

Chapter 4   Seeds and Plants

1. Byrd, *London Diary*, 525.

2. On David Landreth, see *From Seed to Flower* (Philadelphia: Pennsylvania Horticultural Society, 1976), 22; *Pennsylvania Gazette*, Dec. 17, 1751.

3. On Hannah Dubre see *Pennsylvania Gazette*, Nov. 7, 1754; Nov. 20, 1755; Sept. 16, 1756; Mar. 5, 1761; Feb. 2, 1762; Feb. 24, 1763; Sept. 18, 1766; Aug. 4 and Oct. 6, 1768; Feb. 8 and Mar. 1, 1770; Sept. 26, 1771; Feb. 6, 1772; Mar. 16 and Oct. 19, 1774; and July 5, 1775.

4. *Maryland Gazette or the Baltimore General Advertiser*, Dec. 6, 1785; *Maryland Journal*, Jan. 24, 1786.

5. *Virginia Gazette and Weekly Advertiser*, Jan. 25 and 30, 1793.

6. *Virginia Gazetter and General Advertiser Extraordinary*, Mar. 30, 1791.

7. *Virginia Gazette and General Advertiser*, Mar. 7, 1792.

8. Broadside dated Jan. 24, 1793, Manuscripts Collection, National Agricultural Library, Beltsville, Md.

9. *Norfolk Herald*, Sept. 8, 1801.

10. *Virginia Herald*, Feb. 17, 1798, and Jan. 11, 1799.

11. Williamsburg City Land Books for 1782–1805, Department of Historical Research, Colonial Williamsburg Foundation.

12. Kent Brinkley, "Plantsmen and Tradesmen of the Soil," Mar. 27, 1991, unpaginated typescript, Colonial Williamsburg Foundation.

13. *Virginia Gazette*, Sept. 15, 1775.

14. *Virginia Gazette and General Advertiser*, Jan. 8, 1799.

15. Webb-Prentis Papers, Box 2, Alderman Library Collection, University of Virginia, Charlottesville.

16. Personal communication between author and Kent Brinkley, Colonial Williamsburg Foundation.

17. *Virginia Gazette and General Advertiser*, Nov. 12, 1799; *Maryland Gazette*, Mar. 31, 1803.

18. *Federal Gazette and Baltimore Daily Advertiser*, Oct. 29, 1800.

19. *Virginia Gazette and General Advertiser*, Oct. 30, 1801; *Examiner*, Nov. 6, 1801.

20. *Petersburg Intelligencer*, Jan. 21, 1803.

21. Williamsburg City Land Books for 1782–1805 and Williamsburg Personal Property Tax Lists, Colonial Williamsburg Foundation.

22. *Richmond Enquirer*, Nov. 4, 1804; John Wickham to Joseph Prentis, Dec. 15, 1807, Webb-Prentis Papers; *Virginia Gazette and General Advertiser*, Nov. 1, 1806.

23. *Maryland Journal and Baltimore Advertiser*, Mar. 20, 1787; *Maryland Gazette*, Nov. 18, 1790; Walter Family Files, Filing Case "A," MHS; *Federal Gazette and Baltimore Daily Advertiser*, Apr. 29, 1807; *Warner and Hanna Baltimore City Directory*, 1801; *C. W. Stafford Baltimore City Directory*, 1803; James Robinson, *Baltimore City Directory*, 1804.

24. *Maryland Journal and Baltimore Advertiser*, Mar. 30 and Apr. 2, 1790.

25. Faris's Diary: May 14, 1793; May 20, 1796; Apr. 20, 1797; Parkinson, *A Tour in America*, 342.

26. *Maryland Journal and Baltimore Advertiser*, Sept. 23 and 30, 1791.

27. *Maryland Journal and Baltimore Universal Daily Advertiser*, Feb. 9, 1795.

28. Parkinson, *A Tour in America*, 488.

29. *Federal Gazette and Baltimore Daily Advertiser,* Dec. 20, 1803; Apr. 29, 1807; May 9, 1808; Nov. 24, 1808; C. W. Stafford, *Baltimore City Directory,* 1803.

30. H. Lewis, *History of Baltimore* (Baltimore, 1912), 636; Baltimore County Inventories, liber 30, folio 38, MSA.

31. *Federal Gazette and Baltimore Daily Advertiser,* Dec. 20, 1803; Sept. 10, 1808; Oct. 16, 1815; Oct. 4, 1819; Apr. 13, 1820; Sept. 8, 1820.

32. Ibid., Nov. 3, 1802 and Sept. 3, 1808; *American and Commercial Daily Advertiser,* Mar. 1, 1820; *Baltimore County Inventories,* liber 48, folio 586, MSA.

33. *Baltimore Daily Repository,* Apr. 30, 1793.

34. *Baltimore Daily Intelligencer,* May 4, 1794.

35. *Federal Intelligencer and Baltimore Daily Gazette,* Mar. 2, 1795.

36. Ibid.

37. Lawrie Todd, ed., *Life and Writings of Grant Thorburn: Prepared by Himself* (New York: Edward Walker, 1852), 61–66, 93–94.

38. *Federal Intelligencer and Baltimore Daily Gazette,* Mar. 2, 1795.

39. Parkinson, *A Tour in America,* 488; *Federal Gazette and Baltimore Daily Advertiser,* Sept. 23, 1799.

40. William Booth Catalogue, Rare Books Collection, MHS.

41. J. T. Scharf, *History of Baltimore City and County* (Baltimore, 1881), 771.

42. *Baltimore County Inventories,* liber 31, folios 311–15, MSA; *Federal Gazette and Baltimore Daily Advertiser,* Oct. 2, 1819; Samuel Jackson's 1819 *Baltimore City Directory;* C. Keenan, *Baltimore City Directory 1822; American & Commercial Daily Advertiser,* Feb. 25, 1820.

43. Carroll Papers, MS 206: Charles Carroll of Annapolis to CC of Carrollton, Mar. 30, 1761, and Nov. 3, 1770; CC of Carrollton to Christopher Bird, Oct. 13, 1766; CC of Annapolis to CC of Carrollton, July 27, 1775.

44. Lloyd Papers, Manuscripts Collection, MHS, MS 2001, Jan. 12, 1806.

Chapter 5   Laborers

1. Brinkley, "Plantsmen and Tradesmen," n.p.

2. Documents 6.1 and 6.2, box 3, folder 48, Archives of the Maryland Province of the Society of Jesus, Special Collections, Lauinger Library, Georgetown University, Washington D.C.

3. Pennsylvania Gazette, Aug. 23, 1750, and Jan. 28, 1752; entry for June 18, 1751, in John Blair's Diary, Alderman Library Collections, University of Virginia, Charlottesville.

4. Carroll Papers, MS 206, Charles Carroll of Annapolis to CC of Carrollton, Aug. 17 and 27, 1770; *Maryland Journal and Baltimore Advertiser,* Aug. 31, 1784.

5. Peale Diary, vol. 20, p. 53; Carroll Papers: MS 211, p. 10; MS 220, Slave Inventory 1773; Miscellaneous "list of negroes-1781"; *Maryland Gazette:* May 26, 1730; Nov. 29, 1759; June 14, 1770.

6. *Maryland Gazette,* July 1766.

7. Working professional gardeners, for the sake of this discussion, are those designated as "gardener" in Maryland's written records. Garden laborers and helpers include persons

providing semiskilled gardening maintenance tasks. Carroll Papers, MS 206, Charles Carroll of Annapolis to CC of Carrollton, Apr. 10, 1775.

8. MSA/Historic Annapolis Collection, MdHR G-1393, file #6. For white indentured servants in Maryland, also see *Pennsylvania Gazette,* June 20, 1751; Oct. 23, 1753; Dec. 15, 1757; Oct. 23, 1760; July 24, 1766; Nov. 16, 1774; Nov. 15, 1775; June 12, 1776; and Apr. 9. 1777. For gardeners in Maryland before 1804, see Barbara Wells Sarudy, "Eighteenth-Century Gardens of the Chesapeake," *Journal of Garden History* 9, no. 3 (1989): 155–59.

9. Lloyd Papers, Memo Book, 1768–72, box 14: re gardener John Truman, Nov. 1772; re gardener Robert Cushney, Mar. 30, 1772; box 56 in James Holland's Accounts: re gardeners John Bowman and Robert Day.

10. Lloyd Papers, Memo Book 1768–72, box 14, Jan. 23, 1771.

11. Re gardener Anthony Faresteau: Charleston County, S.C., Land Records, Miscellaneous: pt. 63, book B5, Mar. 14, 15, 1779, pp. 348–51; pt. 65, book D5, Mar. 14, 1779, p. 325; pt. 66, book E5, July 23, 1779, pp. 146–49; *Royal South-Carolina Gazette* (Charleston), Sept. 21, 1780; Charleston County, S.C., Renunciation of Dower, transcript, July 22, 1779, pp. 222, 190; ibid., transcript, Mar. 3, 1772, p. 222; *Gazette of the State of South Carolina,* June 9, 1777; *South-Carolina and American General Gazette,* Jan. 19 and Feb. 2, 1776; Charleston County, Wills, vol. 22, 1786–1793, p. 37; Charleston County, Inventories, vol. A, 1783–1787, p. 469.

    Re gardener James Waddell: *Charleston City Gazette,* June 5, 1823; Charleston County, S.C., Letters of Administration, Aug. 13, 1783, p. 282; Charleston County, Land Records Miscellaneous: pt. 70, book M5, Aug. 22, 1783, pp. 421–23; pt. 84, book R6, Aug. 29, 1796, pp. 307–11; pt. 88, book T6, Dec. 12, 1797, pp. 397–99; pt. 89, book W6, Sept. 25, 1798, pp. 277–78; *Charleston City Gazette,* June 5, 1823, pt. 93, book K7, Jan. 1804, pp. 471–72.

12. *Federal Gazette and Baltimore Daily Advertiser,* Mar. 11, 1806; *Maryland Gazette,* July 1766 and Mar. 25, 1803.

13. Carroll Papers, MS 206, Charles Carroll the Barrister's Letterbooks, Jan. 1768.

14. Carroll Papers, MS 206, Charles Carroll of Carrollton to CC of Annapolis, May 27, 1772.

15. Ambler, "Diary," 152–79.

16. Blanche Henry, *British Botanical and Horticultural Literature before 1800* (London: Oxford University Press, 1975), 2:179.

17. *Maryland Gazette,* July 5, 1764.

18. Carroll Papers, MS 206, Charles Carroll of Annapolis to CC of Carrollton, Aug. 31 and Sept. 14, 1770.

19. *Federal Gazette and Baltimore Daily Advertiser,* July 15, 1799; *Federal Intelligencer and Baltimore Daily Gazette,* Feb. 17, 1795; *Maryland Gazette,* July 3, 1751.

20. Carroll Papers, MS 206, Charles Carroll the Barrister Letterbooks, Jan. 1768.

21. Faris's Diary, Mar. 26 and 19, 1792; Apr. 15 and May 8, 1794; Feb. 9, 1796; Sept. 10 and 24, 1797; Feb. 11 and July 22, 1798.

22. Parkinson, *A Tour in America,* 27, 419–21.

23. Ibid.; Faris's Diary, May 10, 1792.

24. Joshua Brooks, "Journal of Joshua Brooks," in *The Icehouses and Their Operations at Mount Vernon,* ed. John P. Riley (Mount Vernon, Va.: Mount Vernon Ladies Association, 1989), 18.

25. Parkinson, *A Tour in America*, 448.

26. Baltimore County Tax List, 1773; *Pennsylvania Gazette*, Sept. 16, 1756; June 29 and July 6, 1769; June 23, 1773; Apr. 10, 1782.

27. Brinkley, "Plantsmen and Tradesmen," n.p.

28. *Federal Gazette and Baltimore Daily Advertiser*, Jan. 4, 1802; *American and Commercial Daily Advertiser*, Oct. 26, 1801 and July 7 1809; *Maryland Gazette or Baltimore Advertiser*, Mar. 20, 1787.

29. Appendixes 2 and 3 in Sarudy, "Eighteenth-Century Gardens of the Chesapeake," 154–57.

30. Ira Berlin, *Slaves without Masters: The Free Negro in the Antebellum South* (New York: Vintage Books, 1976), 23, 46–47. For more on Maryland's free and slave blacks, see Christopher Phillips, *Freedom's Port: The African American Community of Baltimore, 1790–1860* (Urbana: University of Illinois Press, 1997); and T. Stephen Whitman, *The Price of Freedom: Slavery and Manumission in Baltimore and Early National Maryland* (Lexington: University Press of Kentucky, 1997).

31. Suzanne Ellery Greene, *Baltimore: An Illustrated History* (Woodlawn Hills, Calif.: Windsor Publications, 1980), 56. Greene reports 1500 total immigration. University of Maryland professor Gary Browne reports 2600 (conversation with author, Dec. 10, 1986).

32. Bevan, "Gardens in Early Maryland," 251.

33. Landon Carter, *The Diary of Colonel Landon Carter of Sabine Hall, 1752–1778*, ed. Jack P. Greene (Richmond: Virginia Historical Society, 1987), 1088; Parkinson, *A Tour in America*, 27.

34. Ibid., 27, 173, 382.

35. Lloyd Papers, Ledger 7, 1770–91, cashbook, box 50.

36. *Maryland Gazette:* Nov. 22, 1749; Jan. 3, 1750; Nov. 13, 1751; Oct. 12, 1752.

37. *Virginia Gazette*, Nov. 6, 1766; Sept. 28, 1769.

38. York County Virginia Order Book 4 (1774–84), 274; Webb-Prentis Papers, Joseph Prentis to Governor Nelson, Nov. 24, 1781.

39. In Williamsburg, James Hubbard did advertise for a "skillful Gardener" (*Virginia Gazette*, Feb. 17, 1774).

40. *Maryland Journal and Baltimore Advertiser*, Apr. 19, 1783; *Federal Intelligencer and Baltimore Daily Gazette*, Feb. 12, 1795; *Virginia Gazette, or the American Advertiser*, Mar. 5, 1785.

41. *Maryland Herald and Elizabethtown Weekly Advertiser:* July 25, 1799; June 25, 1801; Mar. 30, June 22, July 6, 1803.

42. Luke O'Dio to Thomas Jefferson, June 23, 1801, Papers of Thomas Jefferson, Princeton University Library.

43. Edith Rossiter Bevan, "Gardens in Early Maryland," 261; Parkinson, *A Tour in America*, 227.

44. *Federal Gazette and Baltimore Daily Advertiser*, Jan. 4, 1797; Thompson and Walker, *Baltimore Town and Fells Point Directory*, 1796; Warner and Hanna, *Baltimore City Directory*, 1798–99 and 1801; C. W. Stafford, *Baltimore City Directory*, 1802 and 1803; James Robinson, *Baltimore City Directory*, 1804. In 1796 four gardeners were listed, in 1798–99 three additional gardeners were noted, in 1801 three more. In 1802 eight new gardeners and in 1803 seven joined the ranks, and in 1804 yet another eight gardeners were listed in the directory.

45. Faris's Diary, Mar. 2, 1801.

46. Archives, Baltimore County Orphans Court: WB2, 64 and 121 (for William Lucas); WB 1, 241 (for William Martin).

47. Archives, Baltimore City Orphans Court, Liber 1, 311.

Chapter 6   Garden Books

1. For discussions of eighteenth-century books and libraries in Maryland see: Stuart Sherman, "The Library Company of Baltimore" *Maryland Historical Magazine* 36 (1944): 6–24; and "The Library Company of Baltimore," *Maryland Historical Magazine* 12 (1917): 297–311. Lawrence C. Wroth, *History of Printing in Colonial Maryland, 1686–1776* (Baltimore, 1922). By Joseph T. Wheeler: "Literary Culture in Eighteenth-Century Maryland, 1700–1776" (Ph.D. diss., Brown University, 1938); *The Maryland Press, 1777–1790* (Baltimore, 1938); "Booksellers and Circulating Libraries in Colonial Maryland," *Maryland Historical Magazine* 34 (1939): 111–37; "Thomas Bray and the Maryland Parochial Libraries," *Maryland Historical Magazine* 34 (1939): 246–65; "The Layman's Libraries and the Provincial Library," *Maryland Historical Magazine* 35 (1940): 60–73; "Books Owned by Marylanders, 1700–1776," *Maryland Historical Magazine* 35 (1940): 336–53; "Reading Interests of the Professional Classes in Colonial Maryland, 1700–1776," *Maryland Historical Magazine* 36 (1941): 281–301; "Reading Interests of Maryland Planters and Merchants, 1700–1776," *Maryland Historical Magazine* 37 (1942): 291–310. Sarah Pattee Stetson, "American Garden Books Transplanted and Native, before 1807," *William and Mary Quarterly* 3, ser. 3 (1946): 341–69.

2. Maryland inventories are located at the Maryland State Archives in Annapolis. *Baltimore County Inventories*, liber 4, folio 363, Henry Wetherall, Aug. 2, 1730; folio 436, John Stokes, Jan. 22, 1732.

3. *Anne Arundel County Inventories*, MSA, liber 4, folio 204–7, William Bladen's Inventory.

4. Henry, *British Botanical Literature*, 1:189.

5. Wheeler, "Reading Interests of Maryland Planters and Merchants," 27.

6. Henry Callister's letters, Maryland Episcopal Diocesan Library Collections, Baltimore: *Letterbooks* 4:716, 769; HC to William Carmichael, Jan. 6, 1766; HC to Dr. Leith, Apr. 2, 1765.

7. Richardson Wright, *The Story of Gardening* (New York: Garden City Publishing, 1938), 293; Henry, *British Botnical Literature*, 2: 47, 50.

8. Callister's *Letterbooks*, 4:691, Henry Callister to Governor Horatio Sharpe, Nov. 9, 1764.

9. Henry, *British Botanical Literature*, 2:211.

10. Carroll Papers, MS 208, MHS, Charles Carroll, Barrister, *Letterbook*, CC of Carrollton to Jay Stewart, winter 1755–56.

11. Carroll Papers, MS 208, Charles Carroll of Carrollton to William Anderson, July 20, 1767.

12. Henry, *British Botanical Literature*, 2:90–95.

13. Carroll Papers, MS 208, Charles Carroll of Carrollton to William Anderson, Nov. 2, 1765.

14. Carroll Papers, MS 206, *Catalogue of the Library of Charles Carroll of Carrollton* (Baltimore, 1864). Additional information supplied by Sally Mason, associate editor, Carroll Papers; Frederick County Inventories, MSA, Thomas Bacon, May 26, 1768; Henry, *British Botanical Literature*, 2:424–25.

15. Personal communication between author and Sally Mason, Carroll Papers.

16. Ibid.; Henry, *British Botanical Literature*, 2: 470, 397.

17. Personal communication with Sally Mason, Carroll Papers.

18. *Catalogue of the Annapolis Circulating Library, 1783*, Stephen Clark, compiler, Maryland Episcopal Diocesan Archives, Baltimore; Henry, *British Botanical Literature*, 2:363.

19. Catalogue of the Library Company of Baltimore, 1809, MS 80, MHS.

20. Greene, *Baltimore: An Illustrated History*, 52; Sherman, "Library Company," 9, 18.

21. Catalogue of the Library Company of Baltimore.

22. Ibid.; *Pennsylvania Gazette*, Sept. 20, 1759.

23. The March 1772 order books of the Annapolis factorage Wallace, Davidson and Johnson show an order for a copy of Palladio's *The Four Books of Architecture*, edited by Isaac Ware in London in 1649 (Wallace, Davidson & Johnson Account Books, MSA).

24. "God's Controversy with New England." *Proceedings of the Massachusetts Historical Society*, vol. 12.

25. John Adams, *Works*, 3:395.

Chapter 7   Pleasure

1. Lockwood, *Gardens of Colony and State*, 20; *Maryland Gazette*, Mar. 25, 1784; *Virginia Herald:* Oct. 2, 1798; Aug. 5, 1800; Mar. 2, 1802; Mar. 27, 1803; *Pennsylvania Magazine of History and Biography* 12:454–55.

2. A. F. M. Willich and James Mease, *The Domestic Encyclopedia*, 5 vols. (Philadelphia: William Young Birch and Abraham Small, 1803), 4:52.

3. Peter Martin, "Long and Assiduous Endeavors: Gardening in Early Eighteenth-Century Virginia," in Maccubbin and Martin, *British and American Gardens*, 109–11; Jacob Hiltzheimer, *Extracts from the Diary of Jacob Hiltzheimer of Philadelphia*, ed. Jacob Cox Parsons (Philadelphia: William Fell, 1893), 243.

4. John Jones Spooner, "A Topographical Description of the County of Prince George in Virginia, 1793," *Tyler's Quarterly Historical and Genealogical Magazine* 5, no. 1 (1923): 8.

5. William Byrd, *A Journey to the Land of Eden and Other Papers* (New York: Macy-Masius, 1928), 342; Josiah Quincy, journal entry for May 3, 1773, Quincy Papers, MS Box 61, *Journal of Southern Tour*, Massachusetts Historical Soceity, Boston; Parkinson, *A Tour in America*, 477; Christopher Marshall, *Extracts from the Diary of Christopher Marshall*, ed. William Duane (Albany, New York, 1877), 254.

6. George Rogers, "Gardens and Landscapes in Eighteenth-Century South Carolina," in Maccubbin and Martin, *British and American Gardens*, 151.

7. Thomas Jones to Mrs. Jones, Aug. 6, 1728, "Jones Papers," *Virginia Magazine of History and Biography* 24 (1918): 170; William Byrd, "Progress to the Mines," in *The Writings of Colonel William Byrd of Westover* (New York: Doubleday, Page, 1901), 356–57; Mrs. Frances Baylor to John Baylor, May 25, 1770, *Virginia Magazine of History and Biography* 21 (1913): 90; William J. Hinke, ed. and trans., "Report of the Journey of Francis Louis Michel from Berne, Switzerland, to Virginia, October 2, 1701–December 1, 1702," *Virginia Magazine of History and Biography* 24 (1916): 39; Norton Papers, June 12, 1773, Colonial Williamsburg Foundation Library Manuscripts.

8. *South Carolina and American General Gazette*, July 1, 1771; *Horticultural Register*, Jan. 1, 1835.

9. William Bartram, *Travels Through North and South Carolina, Georgia, East and West Georgia, 1791*, ed. Mark Van Doren (New York: Dover, 1928), 72–73.

10. Charles Brockden Brown, *Weiland or the Transformation*, ed. Jay Fliegelman (New York: Penguin Books, 1991), 31.

11. John S. Bassett, ed., *The Writings of Colonel William Byrd* (New York: Doubleday, Page, 1901), 357–58; Fithian, *Journal*, 32, 45.

12. The stroll was at Colonel Spotswood's estate near Germanna, Virginia, recorded in Byrd's journal on Sept. 28, 1732 (Bassett, *William Byrd*, 360).

13. Deborah Pratt, "1762 to Marcia," Deborah Pratt Ruff Notebook, Manuscript Collection, American Antiquarian Society, Worcester, Mass.

14. Carl Bridenbaugh, ed., *Gentleman's Progress: The Itinerarium of Dr. Alexander Hamilton, 1744* (Chapel Hill: University of North Carolina Press for the Institute of Early American History and Culture, 1948), 63, journal entry for June 26, 1744.

15. Sir Augustus John Foster, 1807, Foster Papers, p. 142, Colonial Williamsburg Foundation Library.

16. Karen Madsen, "To Make His Country Smile: William Hamilton's Woodlands," *Arnoldia* 49 (Mar. 1989): 19.

17. Fithian, *Journal*, 173, 44.

18. Kalm, *Travels*, 341.

19. Robert Hunter, *Quebec to Carolina in 1785–1786: Being the Travel Diary and Observations of Robert Hunter, Jr. a Young Merchant of London*, ed. Louis B. Wright and Marion Tinling (San Marino, Calif.: Huntington Library, 1943), 231; Fithian, *Journal*, 58; Lucinda Lee, *Journal of a Young Lady of Virginia, 1782*, ed. Emily V. Mason (Baltimore, 1891).

20. *Vestry Book of Christ Church Parish*, Middlesex County, Va., July 21, 1713; *South Carolina Gazette*, Feb. 2, 1734; *Pennsylvania Gazette*, Oct. 28, 1736; *Columbian Herald*, Charleston, S.C., Apr. 12, 1787.

21. Durand of Dauphine, *A Huguenot in Exile in Virginia*, ed. Gilbert Chinard (New York, 1934), 11.

22. Letter from Rosalie Stier Calvert to her father, begun June 1809, Calvert, *Mistress of Riversdale*, 206.

23. Fithian, *Journal*, 183; Nicholas Cresswell, *The Journal of Nicholas Cresswell*, ed. Samuel Thornley (New York: Dial, 1924), 30; Weld, *Travels*, 1:187; Benjamin Henry Latrobe, *The Journal of Latrobe: Being the Notes and Sketches of an Architect, Naturalist and Traveler in the United States from 1796 to 1820* (New York: J. Appleton, 1905), 30.

24. Anne Blair to Martha Braxton, Aug. 1769, "Papers of John Blair," *William and Mary Quarterly*, 1st ser., 16 (1907–8): 174–80; Cresswell, *Journal*, 52–53.

25. Fithian, *Journal*, 93.

26. Letter written in 1743 by Eliza Pinckney to Miss Bartlett of London, Eliza Lucas Pinckney, *The Letterbooks, 1739–1762* (Chapel Hill: University of North Carolina Press, 1972), 62.

27. Fithian, *Journal*, 136.

28. Ibid., 114.

29. Ibid., 134.

30. Fortescue Cuming, *Sketches of a Tour to the Western Country*, (Pittsburgh: Cramer, Spear and Eichbaum, 1810), 217.

31. Eliza Lucas Pinckney, in 1742, describing Pinckney's Wappoo Plantation garden near Charleston, South Carolina, in a letter to a friend. Pinckney, *Letterbooks,* 36; Bridenbaugh, *Alexander Hamilton*, 55.

32. Philadelphia *Gazette of the United States,* May 25, 1791.

Chapter 8   Food

1. Charles Leach, *The "Salad" Vegetable in the Colonial Chesapeake*, National Colonial Farm Research Report No. 22 (Accokeek, Md.: Accokeek Foundation, 1984), 2; Elizabeth B. Pryor, *Exotic Vegetables in the Colonial Chesapeake*, National Colonial Farm Research Report No. 18 (Accokeek, Md., Accokeek Foundation, 1983), 1, 2.

2. Ambler, "Diary," 152–70; Grove, "Virginia in 1732"; Lois Green Carr and Lorena S. Walsh, "Changing Lifestyles and Consumer Behavior in the Colonial Chesapeake," in *Of Consuming Interests: Styles of Life in the Eighteenth Century*, ed. Ronald Hoffman, Cary Carson, and Peter J. Albert (Charlottesville: University Press of Virginia, 1992).

3. Parkinson, *A Tour in America*, 254, 255, 563, 564. There were six shillings in a dollar in late 1790s Maryland.

4. Ibid., 339, 340.

5. Many varieties of cabbage were grown by both George Washington and Thomas Jefferson. In the eighteenth century, cabbages were grown throughout Europe, the tight-headed varieties in the north and the loose-leafed types in the south. Faris grew late winter (diary entry for Mar. 15, 1792), early york or battesey (3/23/92), leaf (8/6/92), red leaf (8/6/92), green break leaf (3/30/96), drumhead (5/9/96), savoy (3/1/96), white leaf (3/6/97), large dutch (3/29/99), early york (3/29/99), sugar loafed (3/21/01), and curled savoy (7/4/04). Brussels sprouts were not widely grown at this time, but Faris planted this member of the cabbage family in his Annapolis garden. "Greens," not specified as to type, also appeared for sale in the kitchen garden section of Baltimore nurseryman William Booth's 1810 seed and plant catalogue. Spinach was also grown by Washington and Jefferson, both of whom also grew colewart, now called collards. Broccoli was planted chiefly for its greens in the eighteenth century. The large flower buds eaten today were only a skimpy part of the plant during this period. Faris-McParlin Account Books (MS 353); Faris's Diary; Ann Leighton, *American Gardens in the Eighteenth Century: "For Use or for Delight."* (Boston: Houghton Mifflin, 1976), 206, 210; Carolyn Jabs, *The Heirloom Gardener* (San Francisco: Sierra Club Books, 1984), 205.

6. Parkinson, *A Tour in America*, 338, 339.

7. Onions (*Allium*) were first mentioned by Faris as the "Onions of Egypt." Rocambole was another name for this particular sort of perennial vegetable which had garlic-tasting bulbets at the top of a long stalk that did not die quickly when harvested. He also planted onions he called Cardomon (diary entry of 5/8/97) and Ocoro or Caffee (4/1/01). Both Washington and Jefferson grew onions. Leighton, *American Gardens*, 206, 210; Parkinson, *A Tour in America*, 338.

8. Carroll Papers, MS 206 MHS, Charles Carroll of Annapolis to CC of Carrollton, Oct. 29, 1773; Leighton, *American Gardens*, 206, 210; Jabs, *The Heirloom Gardener,* 176, 177, and 181; Parkinson, *A Tour in America*, 340, 341. The varieties of peas Faris grew included Tip Upon

Tiptoe (diary entry of 4/1/94) and Algernan (5/7/97). Washington grew peas, and Jefferson grew forty different varieties. Faris also bought peas at market, according to account book and diary entries on June 3, 1797 and May 19, 1798. Beans were also found in the gardens of Washington and Jefferson, the latter growing more than forty different types. Faris grew several varieties of beans, including arbor (4/4/92), white beans "from over the Bay" (4/4/92), flowering (4/4/92), Irish Beauty (4/4/92), bunch (3/19/93), dwarf (3/26/97), yellow (4/7/03), dark purple (4/7/03), and speckled (7/7/03).

9. Cauliflower and cucumbers were also grown by Jefferson. Eighteenth-century cucumbers were often short and curved, with rough yellow skins, not the long, cylindrical glossy dark green varieties common today. "Simlins" (a variant spelling of *cymling*) were the type of squash Faris grew. Squashes and pumpkins are native to the Americas and have been found in early pre-Columbian graves. Thomas Jefferson also planted squash in his garden. Leighton, *American Gardens,* 206, 210; Jabs, *The Heirloom Gardener,* 199, 202.

10. Various kinds of melons were grown by Jefferson. Eighteenth-century varieties were predominately green-fleshed and muted in color. Orange-fleshed melons did not become popular until the middle of the nineteenth century. Faris grew musk, nutmeg (5/11/95), and polynac (5/11/95) melons. Watermelons (*Citrullus vulgaris*) were also grown by Washington and Jefferson. The plant is said to have come from Africa. Carroll Papers, MS 206, Charles Carroll of Annapolis to CC of Carrollton, Apr. 10, 1770; Leighton, *American Gardens,* 206, 210; Jabs, *The Heirloom Gardener,* 203; Parkinson, *A Tour in America,* 613.

11. Pryor, *Exotic Vegetables,* 36.

12. Otho Holland Williams Papers, MS 908, MHS, David Bryan to Otho H. Williams, 1794.

13. Bordley, *Essays and Notes,* 407; "List of Articles Wanted for the Garden" (1784), Lloyd Papers; "Letters of Charles Carroll, Barrister," *Maryland Historical Magazine* 35:202.

14. Corn also was grown by Washington, and Jefferson planted more than fifteen different varieties. Faris grew Canadian (diary entry for Mar. 20, 1797), Tossminas (4/22/99), and Indian (5/7/02). Colonists often referred to Indian corn as maize when writing to foreigners. Faris purchased corn and also potatoes from local farmers J. Weems and W. Clements. He bought turnips and potatoes from the Anne Arundel County farms of Nicholas MacCubbin Carroll and John Hunter. In Faris-McParlin Account Books (MS 353), see payment entries for N. MacC. Carroll (1/30/90, 3/15/90, 1/13/90); John Hunter (11/24 and 25/91); J. Weems (7/20/91); W. Clements (12/8/97 and 12/18/97). Leighton, *American Gardens,* 206, 210; Eddis, *Letters from America,* 131.

15. Parkinson, *A Tour in America,* 331–32.

16. Ibid., 199.

17. Beets were also planted in the gardens of Washington and Jefferson. Beets were originally grown for their greens, long before the Germans began to eat the roots as well. Eighteenth century beets were generally coarse, turnip-shaped, and hairy. The compact ball-shaped beets grown today were developed in the nineteenth century. The sugar beet was white, and the mangelwarzel was red. George Washington grew parsnips, too. Leighton, *American Gardens,* 206, 210; Jabs, *The Heirloom Gardener,* 195; Charles Leach, *Early American "Root" Crops,* National Colonial Farm Research Report No. 9 (Accokeek, Md.: Accokeek Foundation, 1984), 2, 7; Parkinson, *A Tour in America,* 195, 335; *Maryland Gazette or Baltimore Advertiser,* June 19, 1787; Ambler, "Diary," 155–62; Grove, "Virginia in 1732," 32.

Poor-house potatoes actually came from a poor house. The 1802 and 1803 Baltimore city directories noted that the Poor House, "Is situated about half a mile from the head of

Howard Street, North-west of the city, and exhibits a most beautiful rural scene. Meadows, gardens, fruit trees, &c. surround the building, which now affords relief generally to the poor." And the 1801 Warner and Hannah map of Baltimore showed the Poor House with extensive garden plots.

18. Washington and Jefferson also grew turnips. Carroll Papers, MS 206, Charles Carroll of Annapolis to CC of Carrollton, Nov. 11, 1770; Joshua Johnson's Letterbook, 1771–74, MSA, Joshua Johnson to his Annapolis partners, June 22, 1772; Leighton, *American Gardens*, 206, 210; Jabs, *The Heirloom Gardener*, 196; Parkinson, *A Tour in America*, 337, 338.

19. Bordley, *Essays and Notes*, 276.

20. Large-rooted radishes were a staple in Europe during the Middle Ages. One sixteenth-century author described a specimen that weighed forty pounds. These types, known as "keeping radishes," grew long and plump and, like turnips, were stored for winter use. One variety of radish in use in eighteenth-century Maryland may have been the type Faris called Dutch Black (diary entry for Apr. 10, 1797), probably introduced into the Americas by the Spanish. It had a taproot as long as 6 inches, a strong flavor, and startling black skin. Faris also planted salmon-colored (4/7/03), scarlet (4/7/03), and short-topped (4/7/03). Radishes, carrots, and horseradish were grown by both Washington and Jefferson.

The carrot is thought to be a native of Afghanistan and was known by the Romans. Around the seventh century, purple-rooted carrots were noticed in the Middle East, but the vegetable did not become common in Europe until the fourteenth century. By that time a mutation had produced a variety with yellow roots. Another mutation in the seventeenth century resulted in carrots with orange roots. By 1620, Dutch breeders had produced two distinct types of orange carrots—long orange, which had large, long roots and was grown for winter storage, and horn, which had finer flesh and sweeter flavor. Carroll Papers, MS 206, Charles Carroll of Annapolis to CC of Carrollton, Oct. 8, 1772; Leach, *Early American "Root" Crops*, 7; Parkinson, *A Tour in America*, 198, 338; Leighton, *American Gardens*, 206, 210; Jabs, *The Heirloom Gardener*, 188.

21. Parkinson, *A Tour in America*, 198.

22. Asparagus and lettuce were also popular with Washington and with Jefferson, who grew more than fifteen varieties of lettuce. Head lettuce dates only to the Middle Ages. The iceberg variety popular today was not developed until the nineteenth century. Faris grew blue (diary entry for Mar. 19, 1793), large imperial head (3/29/99), large holland leaf (3/29/99), large yellow holland leaf (4/22/99), and red (3/12/1800). Leeks were also grown by Washington and Jefferson. Faris planted the cherry pepper, which was only one type of the peppers cultivated in the eighteenth-century Chesapeake. Jefferson planted "Cayenne pepper" in 1768. Washington also grew peppers in his Virginia garden. Eggplant was sometimes called "Brown Jolly" in Virginia, and Jefferson grew eggplant after 1810 in pots. Leighton, *American Gardens*, 206, 210, 417; Jabs, *The Heirloom Gardener*, 187.

23. William Faris eagerly grew cherries, as did other colonial gardeners, who were pleased to find that the type of cherry tree grown against walls in England could survive in America as freestanding trees. People with carefully designed kitchen gardens, like George Washington, enjoyed espaliering them against walls as well as lining them up independently in orchard rows. Washington and Jefferson were both partial to May dukes, carnations, black hearts, white hearts, and morellos. William Faris did not indicate what type of cherry he grew, but he must have been reasonably successful at growing them, for he bought no cherries at market.

In 1764 Washington was planting a great quantity of "Bergamy" pears in long rows interspersed with "Spanish pears." He also planted "Bury," St. Germain, and orange bergamots. His Spanish pear may have been the Bon Chretien, one of the oldest French pears, known in America as the Bartlett. Jefferson noted planting the seckel (he called it "Sickle") pear in 1807.

As for plums, Faris grew the magnum bonum (red) or egg plum, which had a dark red, acid fruit that ripened in July. It was used chiefly for preserves and dried for the holiday season. Washington and Jefferson both grew the greengage plum. The yellow egg and white magnum bonum were the same plum as the "mogul" that Washington and Jefferson recorded growing. The Orleans plum was also planted at Mount Vernon, and Washington planted a dozen damson plums in 1786. At the same time, his gardener Tobias Lear also planted 177 wild Cherokee plums. Jefferson sent to Paris for quantities of wild plum stones. He ordered the French plum "Brugnol" for Monticello, referring to it as the plum most used for drying. The fruit was used in jams as well. Faris-McParlin Account Books (MS 353), Sept. 5, 1799; Leighton, *American Gardens,* 229, 231, 232, 237–41; Eddis, *Letters from America,* 131–32. For other discussions of fruit growing in the colonial Chesapeake see: Elizabeth B. Pryor, *"Heaven's Favourite Gift": Viticulture in Colonial Maryland, Virginia, and Pennsylvania,* National Colonial Farm Research Report No. 23 (Accokeek, Md.: Accokeek Foundation, 1984) and *Orchard Fruits in the Colonial Chesapeake,* National Colonial Farm Research Report No. 14 (Accokeek, Md.: Accokeek Foundation, 1983); Parkinson, *A Tour in America,* 378, 623.

24.   Parkinson, *A Tour in America,* 219, 220, 289. Parkinson observed of one well-to-do landowner who lived north of Baltimore, "Mr. Gough had very great quantities of fruit, particularly apples, which he sold at three dollars and a half the barrel." Harry Dorsey Gough owned Perry Hall, then located about 14 miles north of Baltimore. His estate was planted in orchards and consisted of about 1000 acres. He apparently did not exaggerate the size of his orchard when he called it spacious, for in the *Maryland Journal and Baltimore Advertiser* of March 28, 1790, he offered for sale "2,000 apple trees." The *Maryland Journal,* March 28, 1786, reported that Gough was elected the first president of the Society for the Encouragement and Improvement of Agriculture in Maryland, formed that year.

25.   Carroll Papers, MS 206, Elizabeth Carroll to Charles Carroll of Carrollton, Nov. 30, 1757; CC of Annapolis to CC of Carrollton, Aug. 17 and Oct. 22, 1775.

26.   Eddis, *Letters from America,* 131–32.

27.   Both Washington and Jefferson began as enthusiastic planters of wild Chesapeake grapes, growing them from seeds and cuttings. Later, Washington abandoned his vineyard, but Jefferson continued to plant all sorts of foreign grapes. In 1810, on his garden's eleven uppermost terraces, he planted a grape "grown by Major Adlum of Maryland," from a fruit discovered by William Penn's gardener. Carroll Papers, MS 206, William Graves to Charles Carroll of Annapolis, Jan. 14, 1770; CC of Annapolis to CC of Carrollton, Apr. 20 and Oct. 11, 1770; Apr. 1 and Nov. 26, 1773; Aug. 31, Oct. 12 and 22, 1775; CC of Annapolis to CC of Carrollton, Apr. 3, 1777; CC of Carrollton to CC of Annapolis, May 24 and June 2, 1777; Faris's Diary; Leighton, *American Gardens,* 232–33; [Bordley], *Gleanings,* 338.

28.   Currant bushes (*Ribus*) were also planted by Jefferson and Washington. In 1770 Jefferson listed as "work to be done at Monticello . . . Plant raspberries, gooseberries, strawberries and currants." In 1782 he noted that a quart of currant juice "makes 2 blue teacups of jelly, 1 quart of juice to 4 of puree." His gardener Tobias Lear ordered red, white, and black currants for Mount Vernon. Raisins and dried currants were advertised for sale in Annapolis (*Maryland Gazette,* Nov. 1, 1791) by a Lewis Neth and by a shopowner on Calvert Street

in Baltimore, Joseph Pilgrim (*Baltimore Daily Intelligencer*, Nov. 11, 1793). Leighton, *American Gardens*, 231.

29. *Maryland Gazette or Baltimore General Advertiser*, Jan. 9 and Apr. 9, 1784; June 21, 1785; Aug. 13 and 17, 1790; *Baltimore Daily Intelligencer*, Apr. 12, 1794.

   William Stenson and John Richardson operated grocery stores in Baltimore during the period, where they stocked imported fruits such as lemons, limes, oranges, pineapples, raisins, dried prunes, and citrons. Baltimore grocers Joseph Pilgrim and Nathaniel Peck also advertised these fruits for sale.

30. Bevan, "Gardens in Early Maryland," 251; Ambler, "Diary," 166. For an thorough discussion of the gardens and greenhouse at the home of Charles Carroll the Barrister, see Michael F. Trostel, *Mount Clare* (Baltimore: National Society of Colonial Dames of America in the State of Maryland, 1981.)

31. Parkinson, *A Tour in America*, 613.

32. Gooseberries were among the earliest of Jefferson's fruit fancies. In 1767 he planted twelve cuttings of gooseberries with his almonds and "muscle plumbs." Leighton, *American Gardens*, 231, 481. For a discussion of berries see Charles Leach, *Colonial Berries: Small Fruits Adapted to American Agriculture*, National Colonial Farm Research Report No. 11 (Accokeek, Md.: Accokeek Foundation, 1983).

33. The horse chestnut (*Aesculus*) traveled from Constantinople to England in the sixteenth century. It was so named because its nuts were reputedly good for short-winded horses. George Washington brought nuts of the *A. octandra* variety from Virginia's Cheat River in 1785. Later he ordered three from John Bartram in 1792, as did Jefferson, who ordered "Aesculus virginica, yellow horse chestnut," to be planted at Monticello. During this period, the almond tree was grown chiefly for its flowers and therefore set in a garden not in an orchard. Jefferson grew many nuts experimentally, budding English walnuts on American black walnuts and sending to Paris for gallons of hickory nuts. The hickory family was large; nearly all of it was planted by Jefferson. Washington had a row of shellbark hickory. Both of these famous gardeners planted quantities of pecan nuts, which Washington called Mississippi nuts. "Fill buds" (perhaps filberts) were planted by both, as were both Spanish and American chestnuts and almonds. Jefferson even tried pistachio nuts, and Washington planted a "physic nut" from the West Indies. Faris, as we saw in Chapter 1, grew walnuts, and his account books record that he bought chestnuts at market on October 12 and 14, 1798 and October 29, 1799. Leighton, *American Gardens*, 225, 236; Parkinson, *A Tour in America*, 374.

34. Bergamot balm, or bee balm, is also called Oswego tea; the settlers of Oswego, New York, learned its use from the Indians. Bee balm grew in Thomas Jefferson's garden as well. Catnip was rare in early American gardens. Garlic appeared for sale in the 1810 catalogue of Baltimore nurseryman William Booth. Ginger was apparently grown by few colonists. Indian nutmeg was called "fennel flower," "Roman coriander," and "love-in-a-mist." Jefferson sowed the nutmeg plant in one of his oval beds. Jefferson also grew mint, parsley, sage, thyme, and rosemary. Saffron was apparently grown by few colonists, although it was widely used in Europe, both medicinally and culinarily. Sage, a small, fragrant evergreen herb, was used as an ornamental as well as culinary and medicinal plant. Leighton, *American Gardens*, 210, 460; Brian Halliwell, *Old Garden Flowers* (London: Bishopsgate, 1987), 123; Raymond L. Taylor, *Plants of Colonial Days* (Williamsburg, Va.: Colonial Williamsburg Foundation, 1968), 15.

Chapter 9 Society

1. David John Jeremy, ed., *Henry Wansey and His American Journal of 1794* (Philadelphia: American Philosophical Society, 1970), 123.

2. For Philadelphia public gardens, see Harold D. Eberlein and Van D. Hubbard Cortlandt, "The American 'Vauxhall' of the Federal Era," *Pennsylvania Magazine of History and Biography* 68 (1944): 150–73; and F. H. Shelton, "Springs and Spas of Old-Time Philadelphians," *Pennsylvania Magazine of History and Biography* 47 (1923): 196–237.

3. William Priest, *Travels in the United States of America, Commencing in the Year 1793 and Ending in 1797* (London, 1802), 34.

4. *Virginia Argus,* June 7, 1799.

5. *Pennsylvania Mercury,* May 12, 1737.

6. *Norfolk Gazette and Public Ledger,* Apr. 7, 1809.

7. Eberlein and Hubbard Cortlandt, "The American 'Vauxhall,'" 166; *Aurora and General Advertiser,* July 7, 1795.

8. *Virginia Argus,* July 24, 1805.

9. *Maryland Journal and Baltimore Advertiser,* Apr. 27, 1784.

10. *Norfolk Gazette and Publik Ledger,* May 24, 1805.

11. *Virginia Gazette and General Advertiser,* Oct. 23, 1801.

12. Eberlein and Hubbard Cortlandt, "The American 'Vauxhall,'" 167.

13. Parkinson, *A Tour in America,* 594.

14. Lockwood, *Gardens of Colony and State,* 121.

15. *Baltimore Daily Repository,* May 7, 1793.

16. *Virginia Gazette and General Advertiser,* May 27, 1800.

17. *Virginia Argus,* June 7, 1799.

18. Carroll Papers, MS 206, Charles Carroll of Annapolis to CC of Carrollton, July 14, 1760.

19. For a history of pleasure gardens in England, see Warwick Wroth, *The London Pleasure Gardens of the Eighteenth Century* (London: Macmillan, 1896); J. G. Southworth, *Vauxhall Gardens* (New York: Columbia University Press, 1941); W. S. Scott, *Green Retreats: The Story of Vauxhall Gardens 1661–1859* (London: Odhams Press, 1955); David Coke, *The Muse's Bower Vauxhall Gardens 1728–1786* (Sudbury: Gainsborough's House exhibition catalogue, 1978); John Dixon Hunt, *Vauxhall and London's Saxon Theater* (Cambridge: Chadwyck Hoakes, 1984).

20. Quoted in T. J. Edelstein, *Vauxhall Gardens,* exhibition catalogue (New Haven: Yale Center for British Art, 1983), 47.

21. Scott, *Green Retreats,* 27.

22. Fithian, *Journal,* 106.

23. Warner and Hanna, *Baltimore City Directory,* 1801; C. W. Stafford, *Baltimore City Directory,* 1802, 1803; James Robinson, *Baltimore City Directory,* 1804. Chatsworth, with its large, walled traditional garden, was built around 1750 by Dr. George Walker, one of Baltimore's earliest physicians. By 1754 Dr. Walker had died, leaving Chatsworth to his only child, his daugh-

ter Agnes (1731–83). Agnes Walker married William Lux, a wealthy ropemaker, who went into partnership with his nephew, Daniel Bowley, in a variety of shipping and wholesaling ventures. Lux was vice-chairman of the Baltimore Committee of Correspondence at the outbreak of the American Revolution. He died in 1778, and his wife died in 1783, at which point Chatsworth passed to their son George, who lived there with his wife, the former Catherine Biddle of Philadelphia. After her death in 1790, the estate was sold to William Cooke (1746–1817), an Annapolis lawyer and a former Loyalist, who had served as the agent of various other Maryland loyalists. Cooke owned Chatsworth until 1804. Between 1790 and 1804, the gardens were leased to Richard Gray and commercially opened to the public. Passano Files, Maryland Historical Society Library; Scott, *Green Retreats*, 27.

24. *Baltimore Daily Intelligencer,* Sept. 18, 1794.

25. Ibid., Aug. 9, 1794.

26. Ibid., July 10, 1794.

27. Warner and Hanna, *Baltimore City Directory,* 1801; C. W. Stafford, *Baltimore City Directory,* 1802, 1803; *Baltimore Daily Repository,* May 22, 1793.

28. *Baltimore Daily Intelligencer,* May 31, June 4, July 7, July 10, and July 14, 1794.

29. Warner and Hanna, *Baltimore City Directory,* 1801; C. W. Stafford, *Baltimore City Directory,* 1802.

30. Quoted in Edelstein, *Vauxhall Gardens,* 45.

31. Warner and Hanna, *Baltimore City Directory,* 1801; C. W. Stafford, *Baltimore City Directory,* 1802, 1803; James Robinson, *Baltimore City Directory,* 1804.

32. Quoted in Edelstein, *Vauxhall Gardens,* 47.

33. Warner and Hanna, *Baltimore City Directory,* 1801; C. W. Stafford, *Baltimore City Directory,* 1802, 1803; Parkinson, *A Tour in America,* 77.

34. *Federal Intelligencer and Baltimore Daily Gazette,* May 5, 1795.

35. Quoted in Edelstein, *Vauxhall Gardens,* 28.

Chapter 10 Inspiration and Expression

1. George Washington to the marquise de Lafayette, Apr. 4, 1784, in *The Writings of George Washington,* ed. J. C. Fitzpatrick, 39 vols. (Washington, D.C.: George Washington Bicentennial Commission, U.S. Congress, 1944) 27:385.

2. Liberty Hyde Bailey, *The Standard Cyclopedia of American Horticulture,* 3 vols. (New York: Macmillan, 1925), 1586.

3. Alan Fusonie and Leila Moran, eds., *Agricultural Literature: Proud Heritage—Future Promise* (Washington, D.C.: U.S. Department of Agriculture, 1975), 111; *Poulson's Daily Advertiser,* Sept. 20, 1816.

4. M'Mahon, *American Gardener's Calendar.* 5.

5. George Washington to William Pearce, Jan. 24, 1795, in Fitzpatrick, *Writings of George Washington,* 34:103; Lloyd Papers, May 16, 1799; *Maryland Gazette,* Mar. 31, 1803.

6. Faris's diary entries about these trades and sales: Alexander Contee Hanson—Apr. 18, May 4, Dec. 4, 1801; Apr. 25, 1796; Apr. 27, May 3, 1797; Apr. 21, May 16, 1799; Apr. 17, 1800. Thomas Harwood—May 8, June 13–14, 1792; John O'Donnell—May 6, June 13–14, 1792;

Upton Scott—May 5, 1792; Charles Wallace—Mar. 23, 1792; Maximillian Heuisler, May 14, 1793; Parkinson, *A Tour in America*, 342, 488.

7. Historically, factors influencing people's arrangement of their immediate external environment have included the natural land features and climate, functional needs, religious beliefs, scientific and technological advances, and emerging philosophical theories. M'Mahon pointed out that, in the history of Western gardens, there had been two basic types of garden design: the formal, whose topography, plant materials, and ornaments were intentionally arranged in a balanced geometrical plan; and the naturalistic, whose topography, plant materials, and ornaments were intentionally arranged as they would be found in nature. M'Mahon, *American Gardener's Calendar*, 62.

8. Ibid., 73.

9. Thomas Jefferson to St. John de Crevecoeur, Jan. 15, 1787, in Boyd, *Papers of Thomas Jefferson*, 11:44; Pinckney, *Letterbook*, 36; Cocke Family Papers, Henry Skipwith to John H. Cocke, Mar. 19, 1813.

10. Parkinson, *A Tour in America*, 608–9; Benjamin H. Latrobe, *The Virginia Journals of Benjamin Henry Latrobe, 1795–1798*, ed. Edward C. Carter III, 3 vols. (New Haven: Yale University Press, 1977), 1:165.

11. *Federal Gazette and Baltimore Daily Advertiser*, Sept. 23, 1799.

12. Elaine G. Breslaw, *The Records of the Tuesday Club of Annapolis 1745–1756*, (Urbana: University of Illinois Press, 1988); Carroll Papers, MS 206, Charles Carroll of Carrollton to William Graves, Mar. 7, 1772; CC of Annapolis to CC of Carrollton, June 1, 1772; MS 203, CC of Carrollton to William Graves, Sept. 15, 1765.

13. John Adams, *Works*, 3:395. Apparently the British were well aware of the disdain early Americans had for their ostentation. When Englishman Francis Baily visited Norfolk in 1796 he noted, "Our rooms agreed with the spirit and disposition of our host—none of those ornamental appendages, or luxurious downy beds, so unbecoming the character of those who call themselves republicans; but everything corresponding to the habits of those who pretend to look with a degree of contempt on the degeneracy of a luxurious age." Francis Baily, *Journal of a Tour in Unsettled Parts of North America in 1796 & 1797*, ed. Jack D. L. Holmes (Carbondale: Southern Illinois University, 1969), 20–21.

14. Ellen G. Miles, *American Paintings of the Eighteenth Century* (Washington, D.C.: National Gallery of Art, 1995), 113–14.

15. Drinker, *Diaries*, Apr. 10, 1796.

16. Carroll Papers, MS 206, Charles Carroll of Carrollton to William Graves, Aug. 15, 1774.

17. M'Mahon, *American Gardener's Calender*, 72.

18. E. P. Thompson, "Patrician Society, Plebeian Culture," *Journal of Social History* 8 (1974): 389; David Jacques, *Georgian Gardens: The Reign of Nature*, (London: B. T. Batsford, 1983), 49.

19. M'Mahon, *American Gardener's Calendar*, 25.

20. *Virginia Gazette*, Jan. 6, 1767. For instance, Capt. John O'Donnell built Canton near Baltimore with oriental architectural elements. Twinning, *Travels in America*, 118; Parkinson, *A Tour in America*, 77.

21. See Williamson, *Polite Landscapes*.

22. Leo Marx, *The Machine in the Garden* (New York: Oxford University Press, 1964), 82; A. Rupert Hall, *The Revolution in Science, 1500–1750* (New York: Longman, 1983), Introduction and Chapter 14.

23. John Locke, *Essay on Human Understanding* (New York: Penguin, 1960); Joseph Addison, *The Works of Joseph Addison*, 3 vols. (New York: Harper & Brothers, 1850), 6:324. "Essay on Nature and Art in the Pleasure of the Imagination," *Spectator*, #414, London, 1712; Marx, *The Machine In the Garden*, 92, 93, 164; Carroll Papers, MS 206, Charles Carroll of Carrollton to William Graves, Aug. 15, 1774.

24. Marx, *The Machine in the Garden*, 110, 112, 128; J. Hector St. John de Crèvecoeur, *Letters from An American Farmer*, letter 3.

25. Thomas Jefferson, *Notes on the State of Virginia*, ed. William Penden (Chapel Hill: University of North Carolina Press, 1955). Thomas Jefferson's library contained George Mason's *Essay on Design In Gardening* (London, 1768). Mason stated, "liberty extends itself to the very fancies of individuals: independency has been as strongly asserted in matters of taste. . . . to this . . . modern improvements in gardening may be chiefly attributed" (26–27). Faris's Diary, July 3, 1801.

26. *1809 Catalogue*, Baltimore Library Company, MS 80, MHS; *Catalogue of the Annapolis Circulating Library*, Episcopal Diocesan Archives.

27. M'Mahon, *American Gardener's Calender*, 5.

28. John McMahon, *Historical View of the Government of Maryland*, 434, n. 39.

29. Carroll Papers, MS 206, Charles Carroll of Carrollton to Edmund Jennings, Oct. 14, 1766.

30. William Gordon to George Washington, Feb. 16, 1786, in John D. Norton and Susanne A. Schrage-Norton, "The Upper Garden at Mount Vernon Estate—Its Past, Present, and Future: A Reflection on Eighteenth-Century Gardening," typescript, Mount Vernon Ladies Association Library, 1985, 119.

31. M'Mahon, *American Gardener's Calender*, 6.

32. Ibid., 65.

33. Carroll Papers, MS 206, William Graves to Charles Carroll of Annapolis, Jan. 14, 1770.

Richard L. Bushman, "American High-Style and Vernacular Culture," in *Colonial British America*, ed. Jack P. Green and J. R. Pole (Baltimore: Johns Hopkins University Press, 1984), 360, 364; Eddis, *Letters from America*, 17, 57, 58; Jonathan Boucher, ed. *Reminiscences of an American Loyalist* (Boston: Houghton Mifflin, 1925), 65, 66; Abbé Claud Robin, *New Travels Through North America* (Philadelphia, 1783), 50; "The Journal of Baron von Closen," *William and Mary Quarterly* (Apr. 1953); *Letters of Members of the Continental Congress*, ed. Edmund C. Brunett, 8 vols. (Washington, D.C.: Carnegie Institution, 1934), vol. 7, letter #521 (David Howell to Jonathan Arnold), #544 (Samuel Dick to Thomas Sinnickson), #570 (Elbridge Gerry to Samuel Holton); C. Varlo, Esq., *The Floating Ideas of Nature* (London, 1796), 2:79, 87; Diary of Noah Webster, Jan. 3, 1786, in *Notes on the Life of Noah Webster*, ed. Emily Ellsworth Ford Skeel, 2 vols. (New York: privately printed, 1912); Jedidiah Morse, *American Universal Geography*, (London, 1793), 467; La Rochefoucault-Liancourt, *Travels*, 3:259. For parties, balls, carriage rides, see Faris's Diary: Nov. 15, 1792; July 4 and Nov. 20, 1794; July 4 and Nov. 24, 1795; Jan. 11, Feb. 4 and 11, May 2 and 28, and July 28, 1796; Feb. 12 and Sept. 20, 1797; Nov. 23, 1798; July 4, Sept. 15, Nov. 21, and Dec. 12, 1799; Jan. 6, 1800; Dec. 1, 1801; Dec. 28, 1802. For French lessons, dancing school, finishing school, see Oct. 15 and Nov. 1, 1792; Jan. 24 and Sept. 12–13, 1793; and Jan. 15, 1796.

34. Eddis, *Letters from America*, Feb. 17, 1772; Parkinson, *A Tour in America*, 19.

35. Nancy Shippen, *Nancy Shippen: Her Journal Book*, ed. Ethel Armes (Philadelphia: J.B. Lippincott, 1935), 194.

36. John R. Stilgoe, *Common Landscape of America, 1580 to 1845* (New Haven: Yale University Press, 1982), 7–12; Roderick Nash, *Wilderness and the American Mind* (New Haven: Yale University Press, 1982); Henry Nash Smith, *The Virgin Land*, (Cambridge: Harvard University Press, 1978); "God's Controversy with New England," *Proceedings of the Massachusetts Historical Society*, vol. 12.

37. Bordley, *Essays and Notes*, 394; *Memoirs of the Philadelphia Society for Promoting Agriculture, Containing Communication on Various Subjects in Husbandry and Rural Affairs* (Philadelphia, 1815). A list of St. Anne's parishioners mentioned in Faris's diary can be found in Barbara Wells Sarudy, "The Gardens and Grounds of an Eighteenth-Century Chesapeake Craftsman" (master's thesis, University of Maryland, 1988).

38. Carroll Papers, MS 206, Charles Carroll of Carrollton to Edmund Jennings, Aug. 9, 1771.

39. Thomas Jefferson referred to the gardens and landscape surrounding Monticello as "the canvas." F. D. Nichols and R. E. Griswold, *Thomas Jefferson, Landscape Architect* (Charlottesville: University Press of Virginia, 1981), 111. For gardening as a fine art see *Thomas Jefferson's Garden Book*, 303–4; James D. Kornwolf, "The Picturesque in the American Garden and Landscape before 1800," *British and American Gardens: Eighteenth Century Life* 8, no. 5. (Jan. 2, 1983): 105; College of William and Mary faculty minutes, Dec. 29, 1779, 280; *Catalogue of the Annapolis Circulating Library*, Episcopal Diocesan Archives.

40. Jefferson, *Garden Book*.

41. Pinckney, *Letterbook*, 181; Calvert, *Mistress of Riversdale*, 117, May 19, 1805.

42. Calvert, *Mistress of Riversdale*, 77–78, Mar. 2, 1804; Pinckney, *Letterbook*, 181; William D. Martin, *The Journal of William D. Martin, A Journey from South Carolina to Connecticut in the Year 1809*, ed. Anna D. Emore (Charlotte, N.C.: Heritage House, 1959).

43. William Bentley, *The Diary of William Bentley, D.D.*, 4 vols. (Gloucester, Mass.: P. Smith, 1962), vol. 1, June 12, 1791.

44. M'Mahon, *American Gardener's Calender*, 76.

45. *Maryland Gazette*, June 18, 1752. For a discussion of early Williamsburg, including plays performed, see *The Present State of Virginia, From Whence is Inferred a Short View of Maryland and North Carolina*, ed. Richard L. Morton (London, 1724; Chapel Hill: University of North Carolina Press, 1956). For a discussion of Atlantic cities that includes Williamsburg and Annapolis, see John Oldmixon, *The British Empire in America* (London, 1741; New York: A. M. Kelley, 1969. For the play-going colonial, see Jay Fliegelman, *Declaring Independence: Jefferson, Natural Language, and the Culture of Performance* (Stanford: Stanford University Press, 1993); and Richard Bushman, *The Refinement of America: Persons, Houses, Cities* (New York: Vintage, 1993).

46. Faris's Diary: Oct. 19, 1792; Jan. 17, 1794; June 16 and 19, 1795; Oct. 4, 1798; July 4 and Aug. 29, 1799; Nov. 12, 1803; July 28, 1804; Faris-McParlin Account Books (MS 353), July 3, 1798. George Washington, according to Faris's Diary, attended the theater in Annapolis on Sept. 24, 25, and 26, 1771. See also John Adams's diary entry for Feb. 17, 1774 in Charles F. Adams II, *World of John Adams* (Boston, 1850), 435. Adams also wrote in his diary of Charles Carroll the Barrister's garden at Mount Clare, on Feb. 23, 1774.

47. *Massachusetts Magazine*, Nov. 1794.

48. George Washington to Sir Edward Newenham, Apr. 20, 1787, in Fitzpatick, *Writings of George Washington,* 29:20; Thomas Jefferson to Charles Willson Peale, Aug. 10, 1811; and Thomas Jefferson to William Short, Nov. 28, 1814, in Jefferson, *Garden Book.*

49. M'Mahon, *American Gardener's Calender,* 6.

50. See, for example, the will of Henry Jernigham, Esq., St. Mary's County, 1773–1775, Maryland Prerogative Court, MSA, vol. 39, Wills Book WD #4, p. 143.

51. Weld, *Travels,* 101. For a discussion of cemeteries, see David Sloane, *The Last Great Necessity: Cemeteries in American History* (Baltimore: Johns Hopkins University Press, 1991).

52. June 23, 1789, Robert Carter Daybook, MS Collection, Colonial Williamsburg Foundation Library, CWF M-1439.2.

53. Fithian, *Journal,* 61; Timothy Dwight, *Travels in New England and New York (1800),* ed. B. Soloman (New Haven: Princeton University Press, 1969), 360.

Postscript

1. Edith Tunis Sale, *Historic Gardens of Virginia,* (Richmond, Va.: Garden Clubs of Virginia, 1923), 191.

2. Fanny Coalter, St. George Tucker's daughter, to John Coalter, Apr. 11, 1810, TR/04/.2/ FB(T), Special Collections, Colonial Williamsburg Foundation Library.

3. See Richard Westmacott, *African-American Gardens and Yards in the Rural South* (Knoxville: University of Tennessee Press, 1992); Grey Gundaker, "Tradition and Innovation in African American Yards," *African Arts* 22 (Apr. 1993); John Vlach, *Back of the Big House: The Architecture of Plantation Slavery* (Chapel Hill: University of North Carolina Press, 1993); and Mechal Sobel, *The World They Made Together: Black and White Values in Eighteenth-Century Virginia* (Princeton: Princeton University Press, 1987).

4. Landon Carter, *The Diary of Colonel Landon Carter of Sabine Hall, 1752–1778,* ed. Jack P. Green, 2 vols. (Charlottesville: University Press of Virginia, 1965), 2:1095.

5. The tourist in Virginia has many prospects for exploration (in alphabetical order): Belle Isle in Lancaster County, Belmont in Stafford County, Berkley in Charles City County, Castle Hill in Albemarle County, Cherry Grove in Nansemond County, The Chimneys in Frederick County, Claremont in Surry County, Cleve in King George County, Delk Farm in Smithfield, Elsing Green in King William County, Eppington in Chesterfield County, Fairfield in Clark County, Elmwood and Font Hill in Essex County, Fall Hill and the Falls near Fredericksburg, Federal Hill in Frederick County, Four Mile Tree in Surry County, Gaymont in Caroline County, Grovemont in Richmond County, Hickory Hill in Ashland, Mount Airy in Warsaw, Mount Pleasant and Pleasant Point in Surry County, Olive Hill on the Appomattox River, Salubria in Culpepper County, Shoal Bay in Smithfield, Strawberry Plain in Isle of Wight County, Wigwam in Nansemond County, Wilton in Middlesex County, and Woodbury Forest in Madison County. Special thanks to Alan Brown, Kent Brinkley and Calder Loth who helped compile this list.

6. The best history I've found of change in a garden over time is the insightful chronicle of the Carroll grounds and their owners written by archaeologist Elizabeth Kryder-Reid (Ph.D. diss., Brown University, 1991). She dug in the ground and the primary sources to compile an account that has no equal.

Index

Barbara Wells Sarudy is the executive director of the Maryland Humanities Council and was administrative director of the Maryland Historical Society from 1984 through 1989. She holds a master's degree in history from the University of Maryland at College Park, where she has pursued doctoral study. The landscape consultant for the Historic Charleston restoration team, she is on the board of the Southern Garden History Society and has published in its journal, *Magnolia,* as well as in the *British Journal of Garden History*. She serves on the advisory committee for an encyclopedia of historic American gardening terms which is being prepared for the National Gallery of Art. She sits on several state commissions, including the Maryland Commission on African American History and Culture, and is president of the Coalition for Maryland History and Culture.

p.17   camden peach orchards

         annapolis faris' orchard

p.20   Chatsworth cherry trees

p.31   drawing of parcels of trees (fruit)

p.39   Grundy fruit trees — see 1897 map of location

p.47

p.48

p.50

p.51

p.52   Druid Hill Rows

p.54   Druid Hill plan

p.55

p.58   fruit trees along alley ways

p.62   espaliered trees + openasp

p.116

p.122   moss patches plant fruit trees

p.125

p.143

p.149   garden — where people negotiate ind. position

p.150   garden as sanctuary

p.157

Library of Congress Cataloging-in-Publication Data

Sarudy, Barbara Wells.
    Gardens and gardening in the Chesapeake, 1700–1805 / Barbara Wells Sarudy.
      p.   cm.
    Includes bibliographical references (p.   ) and index.
    ISBN 0-8018-5823-2 (alk. paper)
    1. Gardens—Chesapeake Bay Region (Md. and Va.)—History—18th century. 2. Gardening—
Chesapeake Bay Region (Md. and Va.)—History—18th century. I. Title.
SB451.34.C45S27   1998
712'.09755'18—dc21

                                                                        97-38888
                                                                       CIP